AF357832

Quality Evaluation of Bee Products—Volume II

Quality Evaluation of Bee Products—Volume II

Guest Editors

Liming Wu
Qiangqiang Li

Basel • Beijing • Wuhan • Barcelona • Belgrade • Novi Sad • Cluj • Manchester

Guest Editors

Liming Wu
Institute of Apiculture
Research
Chinese Academy of
Agricultural Sciences
Beijing
China

Qiangqiang Li
Institute of Apiculture
Research
Chinese Academy of
Agricultural Sciences
Beijing
China

Editorial Office
MDPI AG
Grosspeteranlage 5
4052 Basel, Switzerland

This is a reprint of the Special Issue, published open access by the journal *Foods* (ISSN 2304-8158), freely accessible at: www.mdpi.com/journal/foods/special_issues/76T4EO7VO0.

For citation purposes, cite each article independently as indicated on the article page online and using the guide below:

Lastname, A.A.; Lastname, B.B. Article Title. *Journal Name* **Year**, *Volume Number*, Page Range.

ISBN 978-3-7258-4046-5 (Hbk)
ISBN 978-3-7258-4045-8 (PDF)
https://doi.org/10.3390/books978-3-7258-4045-8

Contents

About the Editors

Liming Wu

WU Liming, Professor, Doctoral supervisor, Deputy Director, Institute of Apicultural Research, Chinese Academy of Agricultural Sciences. He is a leading talent in scientific and technological innovation for the Organization Department of the CPC Central Committee, as well as a young and middle-aged scientific and technological innovation leading talent for the Ministry of Science and Technology. Additionally, he serves as a post scientist for the National Bee Industry Technology System, Chief Scientist of the Scientific and Technological Innovation Project "Bee Product Quality and Risk Assessment" Team at the Chinese Academy of Agricultural Sciences, and Group leader of the Apiculture Standardization Working Group within the Chinese Apiculture Society.

He has achieved significant breakthroughs in the areas of bee product production process control, quality, and functional evaluation, resulting in numerous original accomplishments. As a result of his contributions, he was awarded one second prize for national technological invention (as the first winner) and two first prizes at the provincial and ministerial levels (also as the first winner). In the past five years, he has authored or co-authored 61 SCI papers, including 45 in the JCR Q1 area and 15 with an impact factor greater than eight. Additionally, he has been granted twenty-six national invention patents, revised eight national/industry standards, and developed six types of substances that meet national standards.

Qiangqiang Li

LI Qiangqiang, Ph.D., associate professor of the Institute of Apicultural Research, Chinese Academy of Agricultural Sciences, is mainly engaged in the quality control and function evaluation of bee products. She is currently leading one National Natural Science Foundation project and three basic scientific research projects of central public welfare research institutes and contributing to eight important national projects, including the National Key Research and Development project, the National Modern Bee Industrial Technology System project, and several National Natural Science Foundation projects. As the first/corresponding author, she has published more than 30 papers in the *Journal of Advanced Research, Food Chemistry, Environment International, Journal of Agricultural & Food Chemistry*, etc. Additionally, she holds ten valid national patents for inventions and serves as an editor and reviewer of several SCI journals.

Editorial

Bee Products: The Challenges in Quality Control

Qiangqiang Li and Liming Wu *

State Key Laboratory of Resource Insects, Institute of Apicultural Research, Chinese Academy of Agricultural Sciences (CAAS), Beijing 100093, China; liqiangqiang@caas.cn
* Correspondence: apiswu@126.com

1. Introduction

In recent years, there has been a significant surge in demand for unprocessed natural foods due to the growing awareness of consumer health. Consequently, both the food and pharmaceutical industries have displayed considerable interest in exploring alternative options to drugs and functional dietary ingredients derived from bee products [1]. Bee products are increasingly acknowledged and embraced by consumers as natural, environmentally friendly products with essential nutritional and therapeutic value. These include honey, bee pollen, propolis, beeswax, royal jelly, bee larvae, queen embryos, etc. The nutritional and bioactive constituents present in these bee products encompass carbohydrates, proteins, peptides, lipids, vitamins, minerals, polyphenols, carotenoids, terpenes, and trace elements. Extensive research has demonstrated that these active functional ingredients confer remarkable antioxidant, antibacterial, anti-inflammatory, immune-regulating, anticancer, and tumor-inhibiting properties to bee products. The consumption of these bee products as dietary supplements in various forms, such as tablets, capsules, powders, granules, candy bars, oral liquids, etc., is highly recommended. Furthermore, the vast potential of harnessing these diverse bee products extends across various sectors, including the food, pharmaceutical, and cosmetics industries. However, the utilization of these resources faces multiple obstacles due to challenges such as lack of standardization, limitations in toxicological research, regulations regarding safe consumption dosage, etc.

Furthermore, the remarkable mobility of bees and their direct contact with diverse surfaces facilitate the accumulation of environmental pollutants during foraging through inhalation, ingestion, and adhesion to their body hair. Consequently, these pollutants are subsequently transported to the hive [2]. The presence of potential toxic elements in the environment is often reflected in the composition of bee products such as honey, pollen, and propolis [3]. Bee products are influenced by factors like environmental pollution, residual pesticides, industrial activities, etc., resulting in issues related to residual contamination that pose a threat to consumer health. The investigation into resolutions concerning potential chemical and biological contaminants in bee products, encompassing heavy metals, pesticides, antibiotics, pathogenic microorganisms, etc., can facilitate sustainable development within the apiculture industry while concurrently offering guidance on product safety to ensure consumer health protection.

2. Heavy Metal Residues in Bee Products

Due to human activities such as mining, urbanization, agricultural practices, and industrialization, the concentration of chemical pollutants or contaminants in the environment is gradually increasing [4,5]. Bees accumulate heavy metals from the environment through various means such as feeding and body hair. These heavy metals include a range of elements, such as chromium, mercury, manganese, cadmium, lead, arsenic, and silver, among others. Eventually they are transferred to beehives, contaminating bee products like honey, pollen, and propolis [2]. These heavy metals ultimately accumulate or become absorbed into humans and animals, evading complete elimination or degradation, and thereby resulting in adverse health effects [6].

Citation: Li, Q.; Wu, L. Bee Products: The Challenges in Quality Control. *Foods* **2023**, *12*, 3699. https://doi.org/10.3390/foods12193699

Received: 19 September 2023
Accepted: 6 October 2023
Published: 9 October 2023

3. Residues of Agricultural and Veterinary Drugs

The excessive use and improper application of pesticides and veterinary drugs have led to widespread environmental pollution. These pesticide compounds can also appear in beehives, disrupting bee colony development, and significantly impacting the quality of bee products [7]. Additionally, bees are susceptible to diseases such as American or European foulbrood that can be prevented or controlled by antibiotics. However, the overuse and illegal utilization of these veterinary drugs may result in drug residues being transmitted into bee products, potentially posing health risks to consumers. Long-term ingestion of bee products containing antibiotic residues can lead to antibiotic resistance, mutagenesis, teratogenesis, carcinogenesis, etc. [8].

4. Pathogenic Microorganism Contamination

Bee products can be contaminated by various pathogenic microorganisms, directly affecting their safety and quality. Yeasts and a variety of bacteria are present in beehives [9]. Additionally, the intestinal tract of bees contains 70% Gram-negative bacteria (e.g., *Citrobacter*, *Enterobacter*, *Escherichia coli*), 27% Gram-positive bacteria (e.g., *Clostridium* spp., *Streptococcus* spp., *Staphylococcus* spp.), and 1% yeast [10]. The microbial contamination in bee products may originate from beehives, bee intestines, plant flowers, dust, air, etc.; it can also occur during product processing, transportation, and storage processes before being ultimately ingested by consumers [11].

5. Potential Allergens in Bee Products

Allergens are also one of the most potential biological food safety risks in bee products. Due to the presence of plant pollen and bee gland proteins in some bee products, allergens may be present, making it easy to cause allergic reactions [12]. The severity of allergic reactions is closely related to the exposure level of allergens and individual physiological conditions, resulting in symptoms that range from mild coughing to anaphylactic shock. The Allergen Database published by the World Health Organization and International Union of Immunological Societies (WHO/IUIS) (http://allergen.org/, accessed on 1 September 2023) has confirmed and registered twelve major allergens from *Apis mellifera*. To improve the safety of consuming bee products, more allergens from bee products like honey, bee pollen, royal jelly, and bee larvae need further exploration.

6. Traceability and Authenticity Identification of Bee Products

There are counterfeit products present in the bee product market, which may lack compliant production processes and raise concerns about their quality. Counterfeit products have the potential to mislead consumers and disrupt market order while compromising consumer rights. The quality of bee products is influenced by various factors such as the bee farming environment, floral source quality, processing techniques, etc. It is crucial to establish stringent measures for quality control during collection and production processes in order to enhance product quality management and risk control measures that ensure the authenticity and high standards of bee products. Some bee products might be marketed with specific floral sources or ingredients as selling points without accurate testing or verification procedures being conducted. This discrepancy between product labeling and actual content can lead to confusion among consumers. Therefore, it is imperative to conduct qualitative and quantitative analyses of bee products from both geographical and plant sources perspectives in order to trace their origin. This will involve establishing fingerprint profiles, screening characteristic markers for more precise determination of honey's authenticity, fostering consumer trust, and providing crucial support for industry development.

This Special Issue aims to publish quality articles on the "Quality Evaluation of Bee Products". The topics include, but are not limited to, the following:

- Distribution of different nutrients in bee products;
- Residue detection of hazardous substances in bee products;

- Characterization of botanical or geographical markers in bee products;
- Identification of genomic characteristics in bee products;
- Evaluation of biological/functional activities of bee products;
- Application of omic technologies to the composition analysis of bee products.

Author Contributions: Conceptualization, Q.L.; validation, L.W.; investigation, Q.L.; writing—original draft preparation, Q.L.; project administration, L.W.; funding acquisition, L.W. All authors have read and agreed to the published version of the manuscript.

Funding: This work was supported by the Agricultural Science and Technology Innovation Program under Grant (CAAS-ASTIP-2022-IAR).

Conflicts of Interest: The authors declare no conflict of interest.

References

1. Fardet, A. Minimally processed foods are more satiating and less hyperglycemic than ultra-processed foods: A preliminary study with 98 ready-to-eat foods. *Food Funct.* **2016**, *7*, 2338–2346. [CrossRef] [PubMed]
2. Finger, D.; Torres, Y.R.; Quináia, S.P. Propolis as an indicator of environmental contamination by metals. *Bull. Environ. Contam. Toxicol.* **2014**, *92*, 259–264. [CrossRef] [PubMed]
3. Tutun, H.; Aluç, Y.; Kahraman, H.A.; Sevin, S.; Yipel, M.; Ekici, H. The content and health risk assessment of selected elements in bee pollen and propolis from Turkey. *J. Food Compos. Anal.* **2022**, *105*, 104234. [CrossRef]
4. Atamaleki, A.; Sadani, M.; Raoofi, A.; Miri, A.; Bajestani, S.G.; Fakhri, Y.; Heidarinejad, Z.; Khaneghah, A.M. The concentration of potentially toxic elements (PTEs) in eggs: A global systematic review, meta-analysis and probabilistic health risk assessment. *Trends Food Sci. Technol.* **2020**, *95*, 1–9. [CrossRef]
5. Ullah, R.; Jan, F.A.; Gulab, H.; Saleem, S.; Ullah, N. Metals contents in honey, beeswax and bees and human health risk assessment due to consumption of honey: A case study from selected districts in Khyber Pakhtunkhwa, Pakistan. *Arch. Environ. Contam. Toxicol.* **2022**, *82*, 341–354. [CrossRef] [PubMed]
6. Solayman, M.; Islam, M.A.; Paul, S.; Ali, Y.; Khalil, M.I.; Alam, N.; Gan, S.H. Physicochemical properties, minerals, trace elements, and heavy metals in honey of different origins: A comprehensive review. *Compr. Rev. Food Sci. Food Saf.* **2016**, *15*, 219–233. [CrossRef] [PubMed]
7. O'Neal, S.T.; Anderson, T.D.; Wu-Smart, J.Y. Interactions between pesticides and pathogen susceptibility in honey bees. *Curr. Opin. Insect Sci.* **2018**, *26*, 57–62. [CrossRef] [PubMed]
8. Okocha, R.C.; Olatoye, I.O.; Adedeji, O.B. Food safety impacts of antimicrobial use and their residues in aquaculture. *Public Health Rev.* **2018**, *39*, 21. [CrossRef]
9. Vázquez-Quiñones, C.R.; Moreno-Terrazas, R.; Natividad-Bonifacio, I.; Quiñones-Ramírez, E.I.; Vázquez-Salinas, C. Microbiological assessment of honey in Mexico. *Rev. Argent. Microbiol.* **2018**, *50*, 75–80. [CrossRef] [PubMed]
10. Al-Waili, N.; Salom, K.; Al-Ghamdi, A.; Ansari, M.J. Antibiotic, pesticide, and microbial contaminants of honey: Human health hazards. *Sci. World J.* **2012**, *2012*, 930849. [CrossRef]
11. Orantes-Bermejo, F.J.; Pajuelo, A.G.; Megías, M.M.; Fernández-Píñar, C.T. Pesticide residues in beeswax and beebread samples collected from honey bee colonies (*Apis mellifera* L.) in Spain. Possible implications for bee losses. *J. Apic. Res.* **2010**, *49*, 243–250. [CrossRef]
12. Aguiar, R.; Duarte, F.C.; Mendes, A.; Bartolomé, B.; Barbosa, M.P. Anaphylaxis caused by honey: A case report. *Asia Pac. Allergy* **2017**, *7*, 48–50. [CrossRef] [PubMed]

 *foods*

Article

Conventional vs. Organically Produced Honey—Are There Differences in Physicochemical, Nutritional and Sensory Characteristics?

Sladjana P. Stanojević [1,*], Danijel D. Milinčić [1], Nataša Smiljanić [1], Mirjana B. Pešić [1], Nebojša M. Nedić [2], Stefan Kolašinac [3], Biljana Dojčinović [4], Zora Dajić-Stevanović [3] and Aleksandar Ž. Kostić [1]

[1] Department of Food Technology and Biochemistry, Faculty of Agriculture, University of Belgrade, Nemanjina 6, 11080 Belgrade, Serbia; danijel.milincic@agrif.bg.ac.rs (D.D.M.); akismiljanic@gmail.com (N.S.); mpesic@agrif.bg.ac.rs (M.B.P.); akostic@agrif.bg.ac.rs (A.Ž.K.)

[2] Department of Breeding and Reproduction of Domestic and Farmed Animals, Faculty of Agriculture, University of Belgrade, Nemanjina 6, 11080 Belgrade, Serbia; nedicn@agrif.bg.ac.rs

[3] Department of Agrobotany, Faculty of Agriculture, University of Belgrade, Nemanjina 6, 11080 Belgrade, Serbia; stefan.kolasinac@agrif.bg.ac.rs (S.K.); zoradajic@agrif.bg.ac.rs (Z.D.-S.)

[4] Institute of Chemistry, Technology and Metallurgy National Institute of the Republic of Serbia, University of Belgrade, Njegoševa 12, 11000 Belgrade, Serbia; bmatic@chem.bg.ac.rs

[*] Correspondence: sladjas@agrif.bg.ac.rs

Abstract: Honey is a sweet syrup mixture substance produced by honey bees. Contradictory results have been reported on the influence of organic and conventional beekeeping on the properties of honey. The aim of this research was to determine the potential difference between organically and conventionally produced honey of the same botanical origin (linden, acacia, chestnut, meadow). It was shown that the electrical conductivity (0.16–0.98 mS/cm), optical rotation ($-1.00 - (-2.60)$ $[\alpha]_D^{20}$), pH values (3.30–4.95), free acidity (4.0–9.0 mmol/kg), total content of phenolic (76.5–145.9 µg GAE/g dry weight (d.w.)) and flavonoids (48.7–307.0 µg QE/g d.w.), antioxidant potential, phenolic profile, mineral composition, color (-8.62–126.57 mmPfund) and sensory characteristics, although statistically significant differences were found, were not significantly improved better in the organic samples. All organic honey samples were richer in hydroxycinnamic acid derivatives (60.5–112.1 µg CGAE/g d.w.) compared to conventional honey (56.7–91.1 µg CGAE/g d.w.) of the corresponding botanical origin. The results show that organic beekeeping does not lead to the production of honey with significantly better physicochemical, nutritional and sensory properties compared to conventionally produced honey.

Keywords: linden/acacia/chestnut/meadow honey; production method; physicochemical properties; mineral composition; phenolic components; antioxidant potential; Raman spectroscopy; sensory characteristics

Citation: Stanojević, S.P.; Milinčić, D.D.; Smiljanić, N.; Pešić, M.B.; Nedić, N.M.; Kolašinac, S.; Dojčinović, B.; Dajić-Stevanović, Z.; Kostić, A.Ž. Conventional vs. Organically Produced Honey—Are There Differences in Physicochemical, Nutritional and Sensory Characteristics? *Foods* **2024**, *13*, 3573. https://doi.org/10.3390/foods13223573

Academic Editors: Liming Wu, Qiangqiang Li and Lubomir Lapcik

Received: 2 October 2024
Revised: 22 October 2024
Accepted: 5 November 2024
Published: 8 November 2024

1. Introduction

"Honey is the natural sweet substance produced by honey bees from the nectar of plants or from secretions of living parts of plants or excretions of plant-sucking insects on the living parts of plants, which the bees collect, transform by combining with specific substances of their own, deposit, dehydrate, store and leave in the honeycomb to ripen and mature" [1]. Honey is a food product that has been used in human nutrition since ancient times. The earliest evidence of beekeeping by primitive man is painted on the cave walls in Spain, Africa and India, 8000 before the new era. The ancient civilizations of the Egyptians, Greeks and Romans have left traces of the use of honey on monuments, in pyramids and in the works of Greek philosophers and writers [2]. Honey is used not only as a nutritional product but also in traditional medicine and clinical conditions. It has been found that the constituents of honey may have antioxidant, antiproliferative, antimicrobial, anticancer, antimetastatic and antiinflammatory properties [3]. However, if honey is contaminated, it can pose serious risks to human health [4]. In an effort to preserve the quality of honey produced in a traditional way without the use of chemicals in an unpolluted environment, there has been an increase in organic honey production.

Due to the protection of the environment, biodiversity, and human health, the production of organic food is experiencing great growth around the world. Organic honey production requires that the honey-bearing pastures are not exposed to chemicals, pesticides are present in the air/soil or antibiotics, and the bees are not fed with sugar, which is common in conventional beekeeping. Organic beekeeping uses beehives made of natural material that do not pose a risk to the environment or bee products and do not harm the bees. The organic apiary must be located within a radius of 3 km from roads and conventionally cultivated crops. The conversion from a conventional to an organic apiary takes one year [5].

Conventional beekeeping allows the use of a wide range of pesticides for which maximum concentrations in honey are set [6]. In addition, they are prescribed by the values for the maximum allowed concentrations for residues of pharmacologically active substances in honey [7]. Inadequate use of these active substances makes honey a threat to human health.

It is known that the variability of the chemical composition of honey is dependent on many factors, including botanical origin (variety), geographical (place of production) and apiary conditions (including the production system). There are many studies on the differences between honey based on botanical and geographical origin, as well as on agro-technical measures applied in beekeeping and the influence of honey storage conditions or the environment. In contrast, very limited research has been reported on the influence of organic and conventional beekeeping on the characteristics of honey, with conflicting results [8–10]. In addition to the fact that organically produced honey must not contain pesticides and various pharmacological preparations, it is generally assumed that organically-produced honey has a higher nutritional value. But is it really the case? The aim of this research was to determine the potential difference in the physical-chemical, nutritional and sensory characteristics between honey of the same botanical origin produced in organic and conventional beekeeping.

2. Material and Methods

2.1. Sample Collection

Eight samples of honey produced using organic and conventional beekeeping by certified beekeepers were used for the study, differing in their botanical origin (linden, acacia, chestnut, meadow) and their geographical origin from the Balkans (Table S1). These 4 organic and conventional samples have been chosen because they are the most represented on our market. All analytical methods were performed within a maximum of three months, during which time the samples were stored in a freezer, while the sensory analysis was conducted immediately after sample collection. Their freshness at the time of analysis was verified through the determination of hydroxymethylfurfural (<20 mg/kg). The pollen frequency in honey samples was not examined. Plant source was declared by beekeepers who tested honey samples in the certified laboratory (for chestnut honey, 79% and other tested samples >60% of a specific pollen type). All tested samples of organic honey had a certificate from a licensed laboratory that they belonged to the "organic honey" group.

2.2. Analytical Methods

The color and optical density of honey were determined by spectrophotometric measurement of the absorbance (635 nm) of an aqueous honey solution (1:1; w/v) using the Pfund scale after conversion of absorbance values [11]. Other physicochemical parameters were analyzed using IHC methods [12]. The moisture content and soluble solids of the honey samples were determined at 20 °C using an Abbe-type refractometer (Digital Refractometer, Atago Co., Ltd., Tokyo, Japan) and expressed in °Brix. The free acidity of honey is the content of all free acids, determined by the titrimetric method (by titration to pH 8.3) and expressed in millimoles of acid/kg of honey (mmol/kg; [13]). Honey pH value (in 10% aqueous honey solution) and free acidity were determined using pH meter-Consort-C931 (Turnhout, Belgium). The specific optical rotation was measured with an Atago™ Polax-2L polarimeter (Tokyo, Japan) and expressed as the angle of rotation of polarized light at the wavelength of the sodium-D line at 20 °C ($[\alpha]_D^{20}$) of an aqueous honey solution of 1 dm depth containing 1 g/mL of the substance. Electrical conductivity was determined using a 20% (*w/v*) aqueous honey solution at 20 °C (Jenway Conductivity Meter 4310; Stone, UK) and expressed in milliSiemens per centimeter (mS/cm). The diastase activity of the samples was measured using the Phadebas method [12], and the results were expressed in diastase number (DN).

The total phenolic (TPC) and total flavonoids (TFC) content, as well as the content of hydroxycinnamon acid derivatives (DHCA), was determined using spectrophotometric methods (UV-1800, Shimadzu USA Manufacturing Inc, Canby, OR, SAD) according to Kostić et al. [14]. The results were expressed: for TPC as micrograms of gallic acid equivalents per gram of dry sample (μgGAE/g), for TFC as micrograms of quercetin per gram of dry sample (μgQE/g) and for DHCA as micrograms of chlorogenic acid equivalents per gram of dry sample (μgCGAE/g).

The profile of phenolic compounds was analyzed by UHPLC Q-ToF MS analysis in aqueous honey solution (1:2; *w/w*) on an Agilent 1290 Infinity ultra-high-performance liquid chromatography (UHPLC) system coupled with a quadrupole time-of-flight mass spectrometry (6530C Q-ToF-MS) from Agilent Technologies, Inc. Santa Clara, CA, USA, using the same method and operating parameters as previously described in detail by Kostić et al. [14]. Data-dependent acquisition (DDA) was employed for suspect screening, using the Auto MS/MS acquisition mode (100–1700 *m/z*; scan rate 1 spectra/sec), with fixed collision energy set at 30 eV. Agilent MassHunter software (https://www.agilent.com.cn/en/promotions/masshunter-mass-spec) was used for instrument control and data analysis. Phenolic compounds and abscisic acid were quantified using available standards and expressed as μg/g honey. Table S2 shows a list of phenolic compounds used for quantification and their equation parameters. The exact masses of the components were calculated using ChemDraw software (version 12.0, CambridgeSoft, Cambridge, MA, USA).

The antioxidant properties were determined using three methods: DPPH (α,α-diphenyl-2-picrylhydrazyl) radical scavenging activity, ferric-reducing power (FRP), cupric-reducing antioxidant capacity (CUPRAC) by the procedure detailed by Kostić et al. [14,15]. The results were expressed as milligram equivalents of ascorbic acid per gram of dry sample (mgAAE/g) for the FRP and CUPRAC assays and as a percentage of radical inhibition for the DPPH assay.

The concentrations of macro and microelements in the solution obtained after total mineralization were measured using inductively coupled plasma with optical emission spectrometry (ICP-OES) on the Thermo Scientific iCAP 6500 Duo ICP instrument (Thermo Fisher Scientific, Cambridge, UK) with iTEVA software (https://iteva.software.informer.com/). Calibration standard solutions were prepared from three certified standards: Multi-Element Plasma Standard Solution 4, Specpure® (Alfa Aesar GmbH & Co. KG, Emmerich am Rhein, Germany); SS-Low Level Elements ICV Stock and ILM 05.2 ICS Stock 1 (VHG Labs, Inc-Part of LGC Standards, Manchester, NH, USA). Quantification was performed in triplicate on emission lines with minimal spectral interference. The relative standard deviation was RSD < 3%, with calibration curve correlation coefficients > 0.99. The limits of detection were LOD = 0.01–0.5 μg/L, and the limits of quantification were LOQ = 0.1–1 μg/L. Quality control (QC) included using two certified reference materials (CRMs): DORM 4 (NRCC, North Bay, ON, Canada) and EPA Method 200.7 LPC Solution (ULTRA Scientific, Manchester, NH, USA), with recovery of measured concentrations ranging from 98% to 103% [16].

Raman spectroscopy analysis was performed on the confocal Raman microspectroscopy Witec Alpha 300R (Dreieich, Germany) using a 785 nm laser with a power of

80 mW, an integration time of 60 s, and an objective with 10× magnification with a resolution of 1.24 cm^{-1} and the total magnification of 10×. Chemometric analysis of spectra was performed by each honey sample was recorded 36 times, and the final matrix was 288 × 1562 (number of spectra x number of variables). The spectral range between 300 cm^{-1} and 1500 cm^{-1} was selected for further analysis since being informative. The baseline correction, spectra normalization, and 2nd order derivative processing were applied to achieve the best discrimination power. After preprocessing, PCA was performed to reduce the number of variables. Quadratic Discriminant Analysis (QDA) was applied to develop a valid model for the classification of all tested samples, i.e., to obtain the separate classification groups corresponding with the total sample number (8 groups in total, i.e., Linden, Acacia, Chestnut and Meadow from organic and conventional production) and to be agreeable with their botanical origin (4 groups in total i.e., Linden, Acacia, Chestnut and Meadow from both production ways taken jointly). For this purpose, the 5 Principal components (PCs) were used. Validation of the model was performed using the Leave-One-Out-Cross-Validation (LOOCV) method. The chemometric analysis was performed using Unscrambler X software (version 10.4).

2.3. Sensory Analysis

The sensory evaluation was carried out by 10 trained expert panelists in two sessions according to the descriptive semi-quantitative method of. Marcazzan et al. [17]. For the evaluation of the overall acceptability, a "hedonic scale" was used with sixty honey consumers in two sessions, with acceptability grades ranging from 1 to 9 [18]. Sensory analyses were conducted in accordance with the Code of Professional Ethics of the University of Belgrade [19]. At the beginning of the sensory examination, all panelists gave written consent to participate, and they were aware they could withdraw from the study at any time, that their responses were confidential, that the responses would be used for scientific purposes, as well as that the participant's data and their answers will not be published without their knowledge. Before sensory evaluation, participants were fully informed about study requirements. The tested samples were safe for consumption.

2.4. Statistical Analysis

The results of the study were expressed as the mean and pooled standard deviation (Pooled std) of three replicates (unless otherwise indicated). For that, Statistica software version 8.0 (StatSoft Co., Tulsa, OK, USA) was used, as well as for the determination of Pearson correlation coefficients and Tukey's test (at $p < 0.05$). Principal component analysis (PCA) was used to determine the possible correlations between the measured objects.

3. Results and Discussion

3.1. Physicochemical Parameters

Physical parameters (such as electrical conductivity and specific optical rotation) are the basic characteristics of honey that are important for its classification; their measurement is comparatively simple, and they provide important information [20]. Optical rotation is a parameter that shows the botanical origin of the honey and indicates adulteration of the honey [21]. The examined samples of honey of different botanical origins (linden, acacia, chestnut, meadow) belong to the group of nectar-honey, which originate from the nectaries of flowers [22,23]. The values of optical rotation depend on the type of sugar and relative proportions of sugars in the honey. The specific rotation of fructose is $-92.4°$, glucose $+52.7°$, sucrose $+66.5°$, maltose $+130.4°$, melezitose $+88.2°$ and erlose $+121.8°$ [24,25]. Since nectar honey is dominated by fructose, which has a negative specific rotation, the total specific rotation of this type of honey is negative [24–26]. Accordingly, all honey samples examined in this study, as they belong to the nectar-honey group, should have negative values for specific optical rotation, which was also registered (for honey obtained by organic beekeeping—from $-0.10\ [\alpha]_D^{20}$ to $-1.83\ [\alpha]_D^{20}$ respectively; for honey obtained by conventional beekeeping—from $-1.00\ [\alpha]_D^{20}$ to $-2.60\ [\alpha]_D^{20}$ respectively; Table 1).

Table 1. Physicochemical characteristics of tested honey samples, their classification by color and overall sensory acceptability [a].

		Physicocheical Characteristics						
Honey Samples		**Specific Optical Rotation $[\alpha]_D^{20}$**	**Electrical Conductivity (mS/cm)**	**Moisture (%)**	**pH**	**Free Acidity (mmol/kg)**	**Diastasis (DN)**	**Soluble Solids (°Brix)**
organic produced	linden	−1.35 [d]	0.80 [c]	18.39 [a]	3.68 [e]	7.0 [c]	29.50 [a]	77.50 [e]
	acacia	−1.83 [c]	0.16 [h]	14.43 [e]	3.41 [f]	4.0 [e]	8.40 [e]	84.25 [a]
	chestnut	−0.81 [f]	0.95 [b]	16.00 [d]	4.34 [b]	6.5 [cd]	16.20 [d]	82.66 [b]
	meadow	−1.00 [e]	0.98 [a]	17.50 [c]	4.11 [c]	9.0 [a]	16.30 [d]	81.00 [c]
conventionally produced	linden	−1.00 [e]	0.72 [d]	18.29 [a]	3.83 [d]	6.0 [d]	17.20 [c]	77.16 [e]
	acacia	−2.60 [a]	0.19 [f]	18.00 [b]	3.30 [g]	4.0 [e]	18.30 [c]	78.83 [d]
	chestnut	−1.25 [d]	0.81 [c]	18.11 [b]	4.95 [a]	4.0 [e]	29.20 [a]	78.50 [d]
	meadow	−2.10 [b]	0.43 [e]	17.60 [c]	3.23 [h]	8.0 [b]	25.90 [b]	81.00 [c]
Pooled std		0.06	0.003	0.07	0.02	0.25	0.44	0.20

		Classification by colour and overall sensory acceptability			
honey samples		mm Pfund *	colour name	optical density	overall acceptability **
organic produced	linden	29.64 [e]	white	0.378	6.2 [a]
	acacia	1.04 [h]	water white	0.0945	7.7 [b]
	chestnut	126.57 [a]	dark amber	-	7.0 [c]
	meadow	76.06 [d]	light amber	1.389	7.2 [d]
conventionally produced	linden	17.00 [f]	extra white	0.189	6.5 [af]
	acacia	−8.62 [g]	water white	0.0945	7.2 [d]
	chestnut	112.08 [b]	amber	3.008	4.1 [e]
	meadow	84.60 [c]	light amber	1.389	6.7 [f]
Pooled std		0.56	/	/	0.03

[a] Means in the same column with different letters are a significant difference according to the *t*-test ($p < 0.05$).
* Pfund Scale—millimeters of the Pfund scale. ** Overall acceptability was the result of the "hedonic scale" of two replicates; n = 2 × 60). Results are shown as mean and pooled standard deviation (Pooled std) of three replicates (unless otherwise specified).

In addition to optical rotation, the electrical conductivity of honey is a parameter used to test the quality and botanical origin of honey [27]. A strong correlation was registered between these values ($r = 0.81$; Table S3). The greater the content of inorganic elements, organic acids, free amino acids, proteins and complex compounds in honey, the greater the electrical conductivity [28]. Since the mineral elements primarily enter the honey with the pollen, their content depends on the predominant pollen in the honey, which indicates its botanical origin [24]. The electrical conductivity values in the examined samples ranged from 0.16 to 0.98 mS/cm. The obtained results are in agreement with the literature data in which values for the electrical conductivity of honey from different origins of 0.15–1.64 mS/cm were recorded [29]. Organic meadow honey had the highest electrical conductivity value (0.98 mS/cm), while organic acacia honey had the lowest electrical conductivity value (0.16 mS/cm; Table 1). With the exception of acacia honey, all other tested samples of organic honey had higher electrical conductivity values than conventionally produced honey of the same origin. (Table 1). Electrical conductivity is defined by a new international standard for honey [1,22,30], replacing data for ash content in honey. According to the standard regulation [1], the electrical conductivity of honeydew and chestnut honey is above 0.8 mS/cm, while the electrical conductivity in nectar-honey is below 0.8 mS/cm. In the tested samples, the values for the electrical conductivity of organically produced meadow honey deviate from the prescribed value for flower honey. Namely, the value was higher than 0.8 (0.98 mS/cm Table 1). This indicates the possibility that the bees collected flower nectar and honeydew [23]. In the literature, values higher than 0.8 mS/cm are given for the electrical conductivity of meadow honey. For example, Živkov-Baloš and co-workers [31] examined eighteen samples of meadow honey from the Vojvodina region (Serbia) and registered an electrical conductivity in the range of 0.08–1.19 mS/cm. The obtained values for electrical conductivity and optical rotation were in the middle range depending on the moisture content of the examined samples ($r = -0.54$ and $r = -0.57$, respectively; Table S3).

Moisture content is the only compositional criterion of honey, which has to be fulfilled in the world honey trade as part of the Honey Standard [26]. Different moisture content was registered between honey samples of the same botanical origin and different methods of production (organic and conventional). In most of the tested honey (except for linden honey), a higher moisture content was registered in conventionally produced samples. The moisture content was determined to be 14.43–18.39% (Table 1) in the tested samples. The permitted moisture content in honey is up to 20%, according to the Rulebook on the quality of honey and other bee products [32]. A maximum value of 21% for the moisture content is according to the regulation of Codex Alimentarius [33], and the same value has been proposed by the European Commission [34] for the new standard. A higher moisture content can lead to the fermentation of the honey and the formation of acetic acid [26,35]; both processes are undesirable. At the same time, lower moisture content can lead to the development of caramelization and Maillard reactions during honey storage [35]. The values for the moisture content in all the tested samples were below the maximum value recommended in these regulations and in agreement with the values reported in the literature for the moisture content of honey of different botanical and geographical origins. For example, Escuredo et al. [36] studied 187 honey samples and registered an average moisture content of 16.9–18.0%. For example, Machado et al. [37] found a moisture content in the range of 14.2–18.0% when analyzing 51 honey samples. Some national beekeeping organizations (e.g., Belgium, Spain, Austria, Germany, Switzerland and Italy) prescribe maximum moisture content values of 17.5–18.5% for special classes of quality honey [26]. The moisture content of honey depends on the season and the degree of maturity of the honey that has been reached in the hive [38] as well as the botanical origin of the honey, the relative humidity in the room and the processing and storage conditions [24]. The moisture content can affect various parameters of honey quality, such as its crystallization quality, viscosity, solubility, taste, and color [36].

The moisture content correlated moderately with diastase enzyme activity in the tested samples ($r = 0.63$; Table S3). Honey contains small amounts of enzymes, the most important of which are diastase, glucose oxidase, invertase, acid phosphatase and catalase [39]. The honey enzymes have been the subject of numerous studies with the aim of distinguishing between natural and artificial honey [40]. Specifically, diastase activity was used to determine the botanical origin of honey [41]. Today, however, diastase is mainly used as a measure of honey freshness, as the activity of this enzyme decreases in mature and heated honey. Namely, diastase is a thermolabile enzyme that breaks down starch and is used as an indicator of the quality and freshness of honey, as it determines the degree of damage to honey caused by heating or improper storage at high temperatures [26]. The diastase activity in the examined samples ranged from 8.40 to 29.50 DN, and a statistically significant difference was found between the diastase activity of all organic and conventional honey samples of the corresponding botanical species (Table 1). The obtained results are in accordance with the regulations which prescribe that the diastase activity in honey should be more than 8 DN [1,22,30]. Conventional honey had a higher diastase activity than organic honey of the corresponding botanical origin, with the exception of organic linden honey (Table 1). Studies have shown that diastase activity in honey of different origins was registered in a wide interval from 0.40 to 22.08 DN [29], while Persano Oddo et al. [39] registered diastase activity from 0.00 to 50.0 DN analyzing 499 honey samples of different botanical and geographical origin.

The pH value of honey is an indicator of the possibility of microorganism growth. The optimal pH value for the growth of microorganisms is 7.2–7.4, while the acceptable pH of honey is 3.2–4.5, which is considered to inhibit the growth of microorganisms in honey [42,43]. According to Bogdanov and associates [24], honey is acidic, with a pH value of 3.5–5.5. In the tested samples, the pH values were between 3.30 and 4.95 and a statistically significant difference in pH values was determined between all organic and conventional honey samples of the corresponding botanical species (Table 1). In the case of conventional chestnut honey, the pH value was outside the range that is considered suitable in terms of antimicrobial activity. The pH values are used to distinguish nectar-honey (low pH values–3.5–4.5) and honeydew (high pH values—4.5–6.5; [4,44]), but, according to Bogdanov and co-workers [24], all tested honey samples (organic and conventional) can be classified as acidic, regardless of the production method.

There was no established dependence ($r = -0.08$; Table S3) between the pH value and the results of testing the free acidity of honey. This is in agreement with literature data [45], according to which the free acidity and pH value of honey are not directly dependent due to the buffering effect of acids and minerals present in honey [44]. The free acidity values of the tested samples ranged from 4.0 to 9.0 mmol/kg, where the free acidity values of organic honey samples (4.0–9.0 mmol/kg) were higher than the values for conventional honey (4.0–8.0 mmol/kg) of the corresponding botanical origin (Table 1). Increased acidity may indicate a higher mineral content in honey [45] and may lead to sugar fermentation, resulting in the formation of organic acids, which affect the taste and microbiological stability of honey [4]. Also, the acidity of honey is affected by the presence of lactones, esters and inorganic ions in honey [43], as well as phenolic acids, vitamin C and proteins, which donate hydrogen ions and contribute to the acidity of honey [46]. However, although slightly higher values for free acidity were registered in organic honey samples, they are still far lower than the maximum allowed value (50 meq/kg; [1]) as well as than values in studies by other authors (for example 17.55–31.83 meq/kg, [42]; or 12.0–134.5 meq/kg, [44]; or 6.45–124.20 meq/kg, [29]). The obtained values for free acidity in this study are in agreement with the relatively wide range of these values obtained by Šarić and associates [47]. These authors determined values for free acidity of acacia honey of 5.0–15.1 mmol/kg, of chestnut honey of 6.0–21.7 mmol/kg and of meadow honey of 7.0–37.7 mmol/kg, depending on the three annual seasons (2003–2005), which confirms that the acidity of honey is determined by the season of honey collection [48]. A strong correlation was found between the free acidity value and the electrical conductivity of the samples ($r = 0.78$; Table S3), which is in agreement with studies emphasizing that the electrical conductivity of honey depends on the content of acids, with higher acid contents in honey trigger higher conductivity [31,49].

Total soluble solids in the tested samples were expressed in Brix degrees. Since the degree of Brix corresponds to 1% of sugar [49], it can be concluded that the examined samples had a sugar content ranging between 77.16% and 84.25% (Table 1), whereby the samples of conventional honey showed slightly lower values than the samples of organic honey, except for meadow honey (which had the same values in the conventional and organic samples). Most of the samples analyzed contained soluble solids in the range of 78.77–316.92 °Brix, which was in agreement with Solayman and co-workers [29] giving an overview of the physicochemical characteristics of about 1000 honey samples from all over the world. Slightly lower values than these were registered in linden honey of organic and conventional production (77.50 and 77.16 °Brix, respectively; Table 1). Honey is a concentrated aqueous solution composed mainly of a mixture of fructose and glucose, but it also contains at least 22 other carbohydrates [29]. As sugars are the main constituents of honey, the physical characteristics, as well as the sweetness, are attributed to the sugar's composition [29].

Honey can be classified by color, a physical property that is very easy to observe and sensory characteristics that are very important to the consumer. The color of honey can range from very light—water white, through amber tones, to very dark, almost black, with possible shades of greenish, light yellow, or reddish [24]. The color of honey can be evaluated by sensory analysis, as well as by physical methods based on visual comparison, using different scales, such as the Pfund-grading [50]. A statistically significant difference in the Pfund values was found for honey color (Table 1) between organic and conventional honey samples of the corresponding botanical species. The Pfund scale values for the tested samples ranged from −8.62 to 126.57 mm Pfund. The color intensity ranged from watery white to dark amber. The most intensely colored honey among the examined samples was organic and conventional chestnut honey, while organic and conventional acacia honey, according to the Pfund scale, was the lightest (Table 1). The color of honey depends on the chemical composition (mineral elements, pollen and phenolic compounds), botanical and geographical origin, as well as on the method of production, agricultural practices, storage temperature and storage time [51,52]. The darkening of honey can occur due to Maillard reactions, fructose caramelization reactions, and reactions to phenolic compounds during honey storage [51,53]. According to Solayman and co-workers [29], darker-colored honey contained more mineral elements compared to lighter-colored honey. In this study, a moderate correlation ($r = 0.66$; Table S3) was found between honey color and the total content of mineral substances. There are conflicting opinions in various studies about the relationship between honey color and the content of phenolic components. For example, Bogdanov and co-workers [26] found that darker honey contains more flavonoids, anthocyanins, tannins and sugar. According to Moniruzzaman et al. [54], a higher Pfund value and color intensity indicate an increased content of phenolic components, in particular flavonoids in honey. On the contrary, Amiot and co-workers [55] pointed out that darker-colored honey contains fewer flavonoids and more phenolic acid derivatives compared to lighter-colored honey. The results obtained in this study showed a strong dependence between the Pfund values and the content of total flavonoids ($r = 0.88$) and hydroxycinnamic acid derivatives ($r = 0.82$), while a moderate dependence was registered between the Pfund values and the content of total phenolic content ($r = 0.63$; Table S3).

3.2. Content of Phenolic Compounds

Phenolic compounds are aromatic phytochemicals with important antioxidant activity in honey. They are natural products of secondary plant metabolism and reach honey through honeybees [56]. Their range in honey is very wide; according to some authors, it ranges from 5 to 1300 mg/kg [57,58], while according to others from 20 to 2400 g/100 g of honey [44].

A statistically significant difference was found in the results for total phenolic content between organic and conventional honey samples of the corresponding botanical species, except for organic and conventional linden honey (Figure 1a). The value of total phenolic content in the analyzed samples ranged from 76.5 to 145.9 µgGAE/g, with the highest content of total phenolics found in conventional meadow honey at 145.9 µgGAE/g, while the lowest content of these compounds was recorded in conventional acacia honey at 76.5 µgGAE/g. Comparing the samples analyzed, conventional honey of different botanical species contained more total phenolics than organic honey, except for acacia honey samples (Figure 1a). The values for the total phenolic content determined in this study do not differ significantly from the published results. Acacia honey has been reported to have a phenolic content of 0.51–0.63 mgGAE/g [59] and of 129.16–341.67 mgGAE/kg [60]. Different values for the content of total phenolics have also been published for honey of other botanical origins: for example, for linden honey from 12.30 to 15.03 mgGAE/100 g [61]; for chestnut honey of 0.12 mgGAE/g [62] and in the range of 487–1134 mgGAE/kg [63] and for meadow honey of 21.3 mgGAE/100 g [64], as well as of 265.1 mgGAE/mL [65]. Polak-Śliwińska and Tańska [10] registered a significantly higher content of phenolics in conventional samples of examined honey than in samples produced by organic beekeeping. These results indicate that the total phenolic content may be much more influenced by factors such as geographical origin, time of honey collection, method and duration of storage, and agrometeorological conditions than by botanical origin or type of beekeeping.

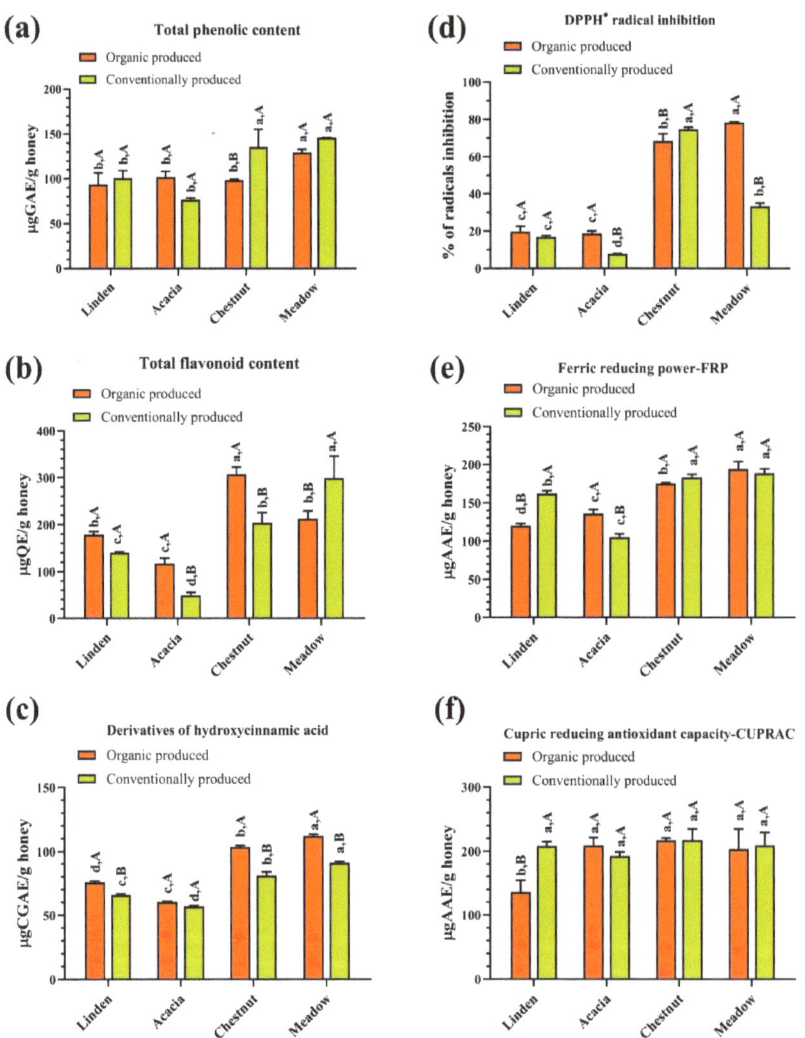

Figure 1. Spectrophotometric assays: (**a**) Total phenolic content; (**b**) Total flavonoid content; (**c**) Total derivatives of hydroxycinnamic acid; Antioxidant assays: (**d**) DPPH radical inhibition activity; (**e**) Ferric reducing power—FRP; (**f**) Cupric reducing antioxidant capacity—CUPRAC. Lowercase letters indicate comparisons of the honey samples of the different botanical origins produced in the same way (organic or conventional); Uppercase letters indicate comparisons of type production of the same botanical origin honey. Different letters indicate statistically significant differences according to Tukey's test ($p < 0.05$).

Flavonoids are a large family of plant phenolic pigments. More than 90% of honey flavonoids come from propolis, suggesting that flavonoids are more important for the identification of geographical origin than in studies on botanical origin [66]. A statistically significant difference was found in the total flavonoid content between organic and conventional honey samples of the same botanical species. The values for the total flavonoid content ranged from 48.7 to 307.0 µgQE/g (Figure 1b). The values of total flavonoids in the organic honey samples were higher than in conventionally produced honey, with the exception of meadow honey (Figure 1b). The organic chestnut honey sample was the darkest of the honey samples analyzed (Table 1) and had the highest total flavonoid content (307.0 µgQE/g), while the conventional acacia honey sample was the lightest and contained the lowest flavonoid level (48.7 µgQE/g; Figure 1b), which is consistent with the data in the literature that darker honey contains more flavonoids and lighter honey contains less [4,54]. Similar to the data for the total phenolic content in honey, the values for the total flavonoid content also vary widely. For example, for acacia and linden honey, the values for flavonoid content range from unidentified [59,62] to the range of 28.83–113.06 mgQE/kg and 20.92–30.32 mgQE/100 g, respectively [60,67]; for meadow honey 6.14 mgQE/100 g [64] or 13.60 mg catechin equivalents/kg [68]. While studies show that flavonoids are mostly not identified in chestnut honey [69], or they are identified at low levels (1.34–3.76 mgQE/100 g; [70]. In this study, almost ten times higher values for the total flavonoid content were found in some samples of chestnut honey analyzed than in these literature data. The reason for this could be the influence of the different geographical origins of the honey.

The levels of hydroxycinnamic acid derivatives were higher in all organic honey samples than in conventionally produced honey. A statistically significant difference was found in the results for hydroxycinnamic acid derivatives content between organic and conventional honey samples of the corresponding botanical species (Figure 1c). The values of hydroxycinnamic acid derivatives ranged from 56.7 to 112.1 µgCGAE/g. The highest content of hydroxycinnamic acid derivatives was found in organic meadow honey at 112.1 µgCGAE/g, while the lowest content of these compounds was in conventional acacia honey at 56.7 µgCGAE/g (Figure 1c). This group of compounds is synthesized by plants via the shikimate metabolic pathway [71] and enters honey via bees, which can contribute significantly to the nutritional value of honey (for example, anticancer, antimicrobial and antioxidant; [72].

3.3. UHPLC Profile of Phenolic Components

A detailed analysis of the profile of bioactive compounds of honey samples should provide several useful pieces of information, such as (1) confirmation of similarities/differences between organically and conventionally produced honey samples; (2) identification of potential markers for the botanical origin of the honey and (3) evaluation of the functional capacity of the analyzed honey. A total of 38 phenolic compounds (in negative ionization mode) and 4 phenylamides (in positive ionization mode) were identified in all analyzed honey samples by UHPLC Q-ToF MS analysis. All compounds were identified based on the exact m/z mass of the molecular ions and the typical MS fragments, as listed in Table 2, while the results of their quantification (µg/g honey) are shown in Table 3. The total content of identified phenolic compounds ranged from 55.62 (conventional chestnut honey) to 1216.91 (conventional meadow honey) µg/g honey. The quantification confirmed that the total content of the identified phenolic compounds originated primarily from propolis-derived flavonoid aglycones (PDFAs) such as pinocembrin, chrysin, pinobanksin and galangin [73] (Table 3), which is consistent with the results of other studies [74–77]. These PDFAs were confirmed in all honey samples (with the exception of galangin in conventional chestnut honey), and their amounts are obviously closely related to the presence and content of propolis in honey. The lowest content of identified total phenolics and PDFAs was found for both chestnut honey (organic and conventional), while their amount varied in the other analyzed honey samples. Furthermore, the total amount of identified phenolics and PDFAs was higher in organic linden, acacia and chestnut honey than in conventional honey, which was not the case for meadow honey. In view of these results, the presence and content of PDFAs are not representative markers for confirming the botanical origin and selecting the production method of honey.

Other phenolic compounds were found selectively and may be potential indicators of the botanical origin of the analyzed honey samples. Among the phenolic acid derivatives, caffeic acid derivatives were the most numerous, but their presence in the honey samples was selective and was below the limit of quantification for most derivatives (<LOQ). The

caffeic acid derivatives were mainly detected in acacia honey as well as in conventional meadow honey. In contrast, these derivatives were completely absent in chestnut and organic meadow honey. The clear differences in the profiles of meadow honey may be due to the presence of different polyfloral pollen grains in the composition of this honey. The detected prenyl caffeate and caffeic acid phenethyl esters (CAPE) are the most common compounds found in propolis [78–80], and their content probably depends on the presence of propolis in honey. Special attention should be paid to ethyl caffeate, which was only detected in linden honey. Benzoic acid, coumaric acid and esculetin were selectively detected in small amounts or traces in the acacia, chestnut and meadow honey examined. Abscisic acid was quantified in both acacia and linden honey as well as in organic chestnut honey. This result is consistent with other studies that have identified abscisic acid as a potential marker for linden and acacia honey [74,76,77].

The phenylamides identified in the honey samples originate from pollen and may be potential botanical origin markers. For example, dicoumaroyl spermidine was only detected in meadow honey, while dicoumaroyl caffeoyl spermidine was only confirmed in chestnut honey. However, further investigations are needed as different coumaroyl derivatives are present in most cases in different bee-collected pollen samples [14,81,82]. Tri-coumaroyl spermidine was found in acacia, chestnut and meadow honey, while phenylamides were not detected in linden honey.

Table 2. Identification and characterization of phenolic compounds in various organic and conventional produced honey samples, using UHPLC Q-ToF MS analysis. Identified compounds, expected retention time (RT), molecular formula, calculated mass, m/z exact mass and MS fragments are presented in Table.

No.	RT	Compound Name	Formula	Calculated Mass	m/z Exact Mass	mDa	MS Fragments
							Phenolic acid and derivatives
1	6.25	Benzoic acid	$C_7H_5O_2{}^-$	121.029	121.0281	−0.85	/
2	7.41	Coumaric acid	$C_9H_7O_3{}^-$	163.0395	163.0385	−1.02	119.0489(100), 120.052(11), 117.0328(8)
3	6.50	Esculetin	$C_9H_5O_4{}^-$	177.0188	177.0176	−1.18	135.043(100), 134.0352(77), 105.033(16), 133.0277(12), 117.0327(9), 121.0276(5), 149.0223(4)
4	6.58	Caffeic acid	$C_9H_7O_4{}^-$	179.0344	179.0341	−0.33	135.0425(100), 134.034(80), 107.0486(12), 117.0318(11)
5	9.87	Ethyl caffeate	$C_{11}H_{11}O_4{}^-$	207.0657	207.0654	−0.33	133.0273(100), 135.0429(76), 134.0342(40), 161.0222(21), 179.0365(2)
6	12.00	Caffeic acid prenyl ester (Prenyl caffeate)	$C_{14}H_{15}O_4{}^-$	247.097	247.0961	−0.93	135.0437(100), 133.028(47), 134.0349(31), 161.0225(26), 179.0331(8)
7	12.11	Caffeic acid benzyl ester	$C_{16}H_{13}O_4{}^-$	269.0814	269.0803	−1.08	134.035(100), 133.0272(79), 161.0232(20), 135.0378(4), 106.0403(4)
8	12.51	Caffeic acid phenethyl ester (CAPE)	$C_{17}H_{15}O_4{}^-$	283.097	283.0962	−0.83	135.0431(100), 161.0229(34), 133.0279(29), 134.0354(23), 179.0331(14)
9	13.04	Caffeic acid cinnamyl ester	$C_{18}H_{15}O_4{}^-$	295.097	295.0961	−0.93	134.0354(100), 133.0278(44), 135.0386(11), 106.0413(5), 161.0218(3)
10	5.80	Caffeic acid hexoside is. I	$C_{15}H_{17}O_9{}^-$	341.0873	341.087	−0.26	161.0229(100), 135.043(87), 179.0333(41), 133.0274(15), 134.0354(9)
11	6.37	Caffeic acid hexoside is. II	$C_{15}H_{17}O_9{}^-$	341.0873	341.087	−0.26	135.0434(100), 179.033(72), 161.0227(54), 134.0352(7)
							Non-phenolic compounds
12	9.47	Abscisic acid	$C_{15}H_{19}O_4{}^-$	263.1283	263.1277	−0.63	203.1064(100), 204.1124(60), 122.035(58), 153.0901(43), 136.0512(41), 189.0899(40), 137.0577(29), 164.0811(20), 138.0666(38), 219.1368(14)
							Phenolic acid amides (Phenylamides)
13	7.74	Di-coumaroyl spermidine	$C_{25}H_{32}N_3O_4{}^+$	438.2393	438.2393	0.02	204.101(100), 147.0435(99), 292.2015(31), 205.1047(15), 275.175(11), 218.117(11), 293.2039(7), 438.2371(6), 119.0491(5)
14	9.91	Tri-coumaroyl spermidine	$C_{34}H_{38}N_3O_6{}^+$	584.2761	584.2763	0.24	438.2382(100), 204.1017(42), 439.2411(35), 147.0439(35), 292.2014(31), 275.1753(16), 420.2271(15), 4212235(9), 293.204(7), 205.1046(7), 119.0494(3)
15	9.75	Dicoumaroyl caffeoyl spermidine	$C_{34}H_{38}N_3O_7{}^+$	600.271	600.2715	0.52	438.2382(100), 204.1011(44), 439.2407(35), 454.2333(26), 292.2005(25), 147.043(16), 420.2272(11), 455.2364(11), 275.1767(9), 163.0393(7), 293.205(6)
16	10.42	Tetra-coumaroyl spermidine	$C_{46}H_{51}N_4O_8{}^+$	787.3707	787.3693	−1.39	641.3327(100), 642.3369(57), 643.339(15), 275.1745(13), 623.3225(8), 204.1021(9), 147.0435(5), 478.2727(4), 494.3009(4)

Table 2. *Cont.*

No.	RT	Compound Name	Formula	Calculated Mass	*m/z* Exact Mass	mDa	MS Fragments
				Flavonoids and derivatives			
				Flavonol aglycones and glycosides			
17	12.37	Galangin	$C_{15}H_9O_5{}^-$	269.045	269.045	0	269.0439(100), 169.0647(19), 171.0438(17), 213.0539(14), 143.0489(13), 223.0384(11), 195.0438(10), 197.0591(9), 211.0386(9), 227.0336(7), 269.0436(8)
18	12.25	Galangin-methyl-ether	$C_{16}H_{11}O_5{}^-$	283.0606	283.0605	−0.15	268.0356(100), 269.0372(23), 240.0404(9), 151.0017(7), 239.0333(7), 117.0332(7), 164.0091(4), 211.0392(4)
19	11.37	Kaempferide	$C_{16}H_{11}O_6{}^-$	299.0556	299.0551	−0.46	284.0306(100), 285.0333(21), 256.0355(9), 133.0277(5), 299.0501(5), 255.0296(2), 257.0433(2), 151.0015(4), 107.0141(4)
20	9.68	Quercetin	$C_{15}H_9O_7{}^-$	301.0348	301.0352	0.37	151.0016(100), 121.0273(45), 107.0114(39), 152.0041(12), 178.9955(9), 149.0223(9), 285.0398(7), 257.0645(5), 243.0235(5)
21	10.49	Isorhamnetin	$C_{16}H_{11}O_7{}^-$	315.0505	315.0497	−0.78	300.0245(100), 109.9994(52), 165.989(49), 255.0283(33), 243.0272(26), 271.0222(22), 301.0296(20)
22	11.39	Rhamnetin	$C_{16}H_{11}O_7{}^-$	315.0505	315.0497	−0.78	165.0176(100), 121.0278(62), 300.0261(22), 151.0022(11), 272.0313(5), 271.0263(5)
23	10.95	Quercetin-dimethyl-ether is. I	$C_{17}H_{13}O_7{}^-$	329.0661	329.0654	−0.73	271.0224(100), 299.017(97), 243.0281(82), 257.0448(24), 300.0202(22)
24	11.73	Quercetin-dimethyl-ether is. II	$C_{17}H_{13}O_7{}^-$	329.0661	329.0659	−0.23	299.0168(100), 271.0234(42), 300.0212(19), 314.0415(12), 301.0235(3), 227.0336(2), 243.0289 (3)
25	9.60	Kaempferol-3-*O*-rhamnoside	$C_{21}H_{19}O_{10}{}^-$	431.0978	431.0968	−1.02	285.0374(100), 284.0306(61), 151.0012(45), 257.0426(34), 431.0957(13), 229.0459(2), 213.0526(3)
26	7.64	Kaempferol 3-*O*-(6″-rhamnosyl)-hexoside-7-O-rhamnoside	$C_{33}H_{39}O_{19}{}^-$	739.2086	739.2065	−2.05	593.1479(100), 594.1509(38), 739.2064(13), 285.038(12), 284.0294(10)
				Flavanonol aglycones and derivatives			
27	10.41	Pinobanksin	$C_{15}H_{11}O_5{}^-$	271.0606	271.0612	0.55	197.059(100), 125.0232(74), 253.0493(67), 161.0595(61), 107.0126(50),151.0032(32), 271.0596(31), 124.0151(29), 181.0643(16), 225.0541(22), 209.0587(14), 254.052(15)
28	9.80	Pinobanksin-5-methyl-ether	$C_{16}H_{13}O_5{}^-$	285.0763	285.0763	0.00	252.0411(100), 138.0306(57), 224.0459(55), 241.0493(32), 253.0447(24), 195.0443(18), 213.054(17), 165.0168(14)
29	13.30	Pinobanksin-3-*O*-propionate	$C_{18}H_{15}O_6{}^-$	327.0869	327.0858	−1.06	253.0487(100), 254.0516(21), 209.0589(6), 197.0582(6), 107.012(4), 271.0579(3), 255.054(3), 185.0578(2), 225.0533(2)
30	13.78	Pinobanksin derivative	$C_{19}H_{15}O_6{}^-$	339.0869	339.0866	−0.26	253.0480(100), 254.0506(19), 197.0587(7), 209.0585(6), 143.0481(5), 107.0119(4), 255.0552(2)
31	14.16	Pinobanksin-3-*O*-butyrate	$C_{19}H_{17}O_6{}^-$	341.1025	341.1017	−0.81	253.0486(100), 254.0523(19), 197.059(5), 209.0592(4), 143.0485(3), 107.0121(3), 255.0549(3), 271.0594(2)
32	14.19	Pinobanksin-3-*O*-pentanoate is. I	$C_{20}H_{17}O_6{}^-$	353.1025	353.1009	−1.61	253.0491(100), 254.0517(22), 197.0592(5), 209.0589(5), 143.0487(4), 255.0536(3), 107.0126(3), 185.0587(2)
33	14.90	Pinobanksin-3-*O*-pentanoate is. II	$C_{20}H_{19}O_6{}^-$	355.1182	355.1175	−0.66	253.0487(100), 254.0524(19), 197.0593(5), 209.059(49, 143.0483(3), 255.0541(3), 107.0123(2), 185.0587(1)
34	15.61	Pinobanksin-3-*O*-hexanoate	$C_{21}H_{21}O_6{}^-$	369.1338	369.1324	−1.41	253.0484(100), 254.0514(18), 197.0579(5), 271.0605(3), 209.0601(3), 143.0464(2)
				Flavone aglycones			
35	12.10	Chrysin	$C_{15}H_9O_4{}^-$	253.0501	253.0502	0.12	253.049(100), 143.0486(68), 107.0127(47), 145.0285(24), 151.0024(24), 119.0488(23), 209.0593(20), 171.0439(16), 185.0594(14), 213.0541(12)
36	10.32	Apigenin	$C_{15}H_9O_5{}^-$	269.045	269.0446	−0.4	117.0324(100), 151.0013(41), 107.0117(37), 269.0435(28), 149.0229(23), 197.0584(15), 225.0526(13)
37	11.18	Genkwanin	$C_{16}H_{11}O_5{}^-$	283.0606	283.0609	0.25	211.0383(100), 239.0329(59), 212.0414(16), 240.0375(14), 167.048(3), 268.0345(3), 283.0589(3)
38	12.71	Acacetin	$C_{16}H_{11}O_5{}^-$	283.0606	283.0605	−0.15	211.0384(100), 239.0331(66), 212.0421(17), 240.0382(15), 268.0358(5), 167.0485(3), 241.0404(2), 213.0444(2)
39	10.78	Luteolin-methyl-ether	$C_{16}H_{11}O_6{}^-$	299.0556	299.0547	−0.86	255.0281(100), 227.0331(79), 284.0303(24), 257.0339(3), 211.0373(3), 132.0194(2), 107.0116(1)
				Flvanone aglycones			
40	12.27	Pinocembrin	$C_{15}H_{11}O_4{}^-$	255.0657	255.0666	0.87	107.0132(100), 171.0445(93), 151.0028(88), 145.065(76), 213.055(62), 255.0652(43), 185.0596(38), 211.0748(18)
41	13.11	Pinostrobin	$C_{16}H_{13}O_4{}^-$	269.0814	269.0807	−0.68	121.0275(100), 165.0177(76), 269.0785(58), 227.0688(57), 183.0791(37), 171.0434(45), 150.0311(30)
42	12.18	Sakuranetin	$C_{16}H_{13}O_5{}^-$	285.0763	285.0763	0.00	164.0098(100), 136.0146(75), 108.0201(41), 151.0021(30), 107.0122(25), 243.0643(16), 285.0743(15), 270.0441(12), 165.0141(11), 201.0532(7), 227.0322(5)

Abbreviations: "is."—isomer.

Table 3. Quantification of phenolic compounds (µg/g) identified in various organic and conventional produced honey samples, using UHPLC Q-ToF MS.

No.	Compound Name	Honey (µg/g honey)							
		Organic Produced				Conventional Produced			
		Linden	Acacia	Chestnut	Meadow	Linden	Acacia	Chestnut	Meadow
	Phenolic acid and derivatives								
1	Benzoic acid [b]	-	-	-	-	-	<LOQ	-	-
2	Coumaric acid [b]	-	2.33	-	<LOQ	-	-	<LOQ	1.48
3	Esculetin [b]	-	<LOQ	<LOQ	<LOQ	-	1.37	-	1.95
4	Caffeic acid [a]	-	<LOQ	-	-	-	-	-	-
5	Ethyl caffeate [b]	<LOQ	-	-	-	2.66	-	-	-
6	Caffeic acid prenyl ester (Prenyl caffeate) [b]	-	5.01	-	-	-	2.47	-	1.72
7	Caffeic acid benzyl ester [b]	-	<LOQ	-	-	-	1.43	-	-
8	Caffeic acid phenethyl ester (CAPE) [b]	-	6.28	-	-	-	<LOQ	-	8.35
9	Caffeic acid cinnamyl ester [b]	-	1.98	-	-	-	<LOQ	-	12.03
10	Caffeic acid hexoside is. I [b]	-	<LOQ	-	-	-	<LOQ	-	<LOQ
11	Caffeic acid hexoside is. II [b]	-	2.17	<LOQ	-	-	<LOQ	-	<LOQ
	Non-phenolic compounds								
12	Abscisic acid [b]	3.06	7.98	13.03	<LOQ	1.85	6.36	-	-
	∑ phenolic acid derivatives + abscisic acid	3.06	25.75	13.03	-	4.51	11.63	-	25.54
	Phenolic acid amides (Phenylamides)								
13	Di-coumaroyl spermidine [b]	-	-	-	5.41	-	-	-	7.41
14	Tri-coumaroyl spermidine [b]	-	<LOQ	3.88	2.02	-	<LOQ	5.54	9.09
15	Dicoumaroyl caffeoyl spermidine [b]	-	-	<LOQ	-	-	-	<LOQ	-
16	Tetra-coumaroyl spermidine [b]	-	-	-	-	-	-	-	1.66
	∑	-	-	3.88	7.43	-	-	5.54	18.16
	Flavonoids and derivatives Flavonol aglycones and glycosides								
17	Galangin [c]	80.27	119.72	9.10	15.34	23.47	47.68	-	107.30
18	Galangin-methyl-ether [c]	5.59	5.20	<LOQ	-	-	-	-	-
19	Kaempferide [c]	3.21	-	<LOQ	-	-	<LOQ	-	-
20	Quercetin [a]	-	-	-	-	-	-	-	3.64
21	Isorhamnetin [c]	-	6.38	9.85	-	-	3.95	9.09	6.97
22	Rhamnetin [c]	-	4.21	-	-	-	1.19	-	1.14
23	Quercetin-dimethyl-ether is. I [c]	-	2.33	-	-	-	1.21	-	6.69
24	Quercetin-dimethyl-ether is. II [c]	21.37	25.02	8.59	-	-	-	-	-
25	Kaempferol-3-O-rhamnoside [c]	-	4.56	-	-	-	-	-	-
26	Kaempferol-3-O-(6″-rhamnosyl)hexoside-7-O-rhamnoside [c]	-	6.66	-	-	-	-	-	-
	∑	110.45	174.06	27.54	15.34	23.47	54.03	9.09	125.74
	Flavanonol aglycones and derivatives								
27	Pinobanksin [c]	102.75	124.18	14.04	38.06	40.91	86.17	1.53	160.69
28	Pinobanksin-5-methyl-ether [c]	16.89	53.33	5.68	3.63	9.96	27.76	-	86.76
29	Pinobanksin-3-O-propionate [c]	-	-	-	-	-	-	-	5.15
30	Pinobanksin derivative [c]	-	-	-	-	-	-	-	1.96
31	Pinobanksin-3-O-butyrate [c]	-	-	-	-	-	-	-	18.40
32	Pinobanksin-3-O-pentanoate is. I [c]	2.75	-	-	-	-	-	-	15.16
33	Pinobanksin-3-O-pentanoate is. II [c]	15.40	3.36	-	-	-	-	-	11.99
34	Pinobanksin-3-O-hexanoate [c]	-	-	-	-	-	-	-	5.34
	∑	137.79	180.87	19.72	41.70	50.87	113.92	1.53	305.45
	Flavone aglycones								
35	Chrysin [a]	163.17	144.74	38.64	69.48	78.60	100.80	16.52	157.10
36	Apigenin [a]	-	12.29	-	3.43	-	5.52	-	4.13
37	Genkwanin [d]	-	9.84	-	-	-	4.87	-	30.67
38	Acacetin [d]	50.67	37.43	5.00	13.65	17.28	22.47	-	75.98
39	Luteolin-methyl-ether [d]	-	14.09	-	3.03	-	4.75	-	15.71
	∑	213.84	218.38	43.64	89.59	95.88	138.41	16.52	283.59
	Flvanone aglycones								
40	Pinocembrin [a]	404.47	428.49	69.30	167.91	174.27	289.59	22.95	458.45
41	Pinostrobin [e]	-	-	10.44	-	-	-	-	-
42	Sakuranetin [e]	16.80	-	-	17.29	-	-	-	-
	∑	421.27	428.49	79.74	185.20	174.27	289.59	22.95	458.44
	∑∑	886.40	1027.57	187.55	339.25	349.00	607.58	55.62	1216.91

Abbreviations: Compound content expressed using available standards [a]; Compounds expressed as caffeic acid equivalent [b]; Compounds expressed as quercetin equivalent [c]; Compounds expressed as chrysin equivalent [d]; Compounds expressed as pinocembrin equivalent [e]; <LOQ—less of limit of quantification; "-"—nonidentified/nonquantified phenolic compounds.

Among the flavonols (with the exception of galangin), significant amounts of quercetin dimethyl ether (compound 24, Table 3) were detected in organic linden and acacia honey and isorhamnetin in both chestnut kinds of honey. Two kaempferol rhamnosides (compounds 25 and 26, Table 3) were only detected in organic acacia honey. These compounds originated from the nectar of acacia flowers and are typical markers for this honey [83]. Pinobanksin esters are characteristic compounds from propolis, and their content probably depends on the proportion of propolis in the honey. These pinobanksin derivatives are easy to detect as they show a typical fragmentation with two main fragments at 271 m/z (deprotonated pinobanksin) and 253 m/z (-H$_2$O) (Table 2). Apart from pinobanksin, a significant content of pinobanksin-5-methyl ether (compound 28) was detected in all analyzed honey, except in organic chestnut honey. Other detected pinobanksin derivatives (except pinobanksin-3-O-pentanoates) were only detected in conventional meadow honey. Acacetin was only not detected in conventional chestnut honey, while its content in the other honey samples varied between 5.00 and 75.98 µg/g. Apigenin and luteolin-methylether were found in both acacia and meadow honey, while genkwanin was only confirmed in acacia honey. Sakuranetin was quantified in organic linden and meadow honey.

3.4. Antioxidant Properties

Many of the honey phenolic compounds are known to have antioxidant activity [70,84,85]. Many authors reported that the content of certain phenolic compounds has a strong linear correlation with the antioxidant activity of honey [84–87]. This study showed that antioxidant activity is not strongly dependent on the content of certain phenolic compounds in honey. Indeed, depending on the method used to determine the antioxidant properties of the honey samples tested, a weak, medium, and strong correlation was found (Table S3). This indicates that the antioxidant properties of honey are not entirely due to the phenolic compounds alone. Although individual phenolics may have considerable antioxidant potential, there may be antagonistic or synergistic interactions between non-phenolic and phenolic compounds. The other constituents (e.g., carotenoids, α-tocopherol, ascorbic acid, organic acids, amino acids and proteins, enzymes (glucose-oxidase, catalase), minerals or Maillard reaction products (melanoidins) that are present in raw honey [88,89] could contribute to the overall antioxidant activity. For example, Meda and co-workers [90] found a higher correlation between radical scavenging activity and proline content than with total phenolic compounds. Therefore, in complex food matrices, where there are many potential antioxidants with different mechanisms of action, it is recommended to use several different methods to determine the antioxidant capacity [91]. Three methods were used to evaluate the antioxidant activity of the tested honey samples: DPPH• radical scavenging capacity (DPPH), ferric-reducing power (FRP) assay and cupric-reducing antioxidant capacity (CUPRAC).

The antiradical activity of the honey samples was estimated using the DPPH assay, and stronger activities were observed in organically produced honey than in conventionally produced honey (18.62–78.35 and 7.72–74.68% radical inhibition, respectively; Figure 1d). The chestnut honey samples deviated from this, but no statistically significant differences were observed between the degree of radical inhibition of organic and conventional chestnut honey. The obtained results for radical inhibition differed significantly for the honey of different botanical origins, with the lighter honey (linden and acacia, from organic and conventional beekeeping) showing significantly lower values (7.72–19.72% radical inhibition; Figure 1d). The strong correlation between the hydroxycinnamic acid derivatives content (r = 0.83; Table S3) and the values obtained by the DPPH assay suggests that these phenolic acids play an important role in the inhibition of DPPH• radicals. The ratio obtained between the total phenolic compounds and the degree of radical inactivation (Table S3) was in agreement with the results of [92], who also emphasized the mean correlation between these parameters in the study of 32 honey of different floral origin. The DPPH assay is frequently used to test the antioxidant properties of honey and is mainly used to determine phenolic antioxidants soluble in organic media [93]. Wilczyńska [92] pointed out in her study that linden honey has a radical inhibition of 63.64% and acacia honey of 35.90%, while Predescu and co-workers [94] registered a degree of radical inhibition of 45.12% in meadow honey. Significantly lower values for the degree of free radical inhibition in linden and acacia were registered in the examined honey samples (Figure 1d) compared to the results of Wilczyńska's [92]. This could be due to the different geographical origins of the honey samples.

The FRP assay shows the ability of antioxidants to reduce Fe^{+3} ions, with a higher value indicating a stronger reducing power [93]. Even when using the FRP assay, no general differences were found between organically and conventionally produced honey (105.1–194.4 mgAAE/g; Figure 1e). The darker-colored honey (organically and conventionally produced chestnut and meadow honey; Figure 1e) showed the highest ability to reduce Fe^{+3} ions. The lightest acacia honey showed the lowest values for antioxidant activity, as determined by the FRP assay. Since strong correlations were found between the results of the FRP assay and the content of phenolic compounds (TPC, TFC, DHCA; Table S3), it can be concluded that phenolic compounds play an important role in the reduction of Fe^{+3} ions. In the literature data available to us, there are no studies of the antioxidant properties of honey using the FRP assay.

The results of the antioxidant activity test using the CUPRAC method show that both groups of honey tested (organic and conventional) show a high ability to reduce Cu^{+2} ions (136.1–217.5 mgAAE/g) with the lowest value recorded for the organic meadow honey (Figure 1f). The highest antioxidant activity was found for chestnut honey, with no significant differences between chestnut honey samples from organic and conventional beekeeping (216.9 and 217.5 mgAAE/g, respectively; Figure 1f). In fact, the results obtained with the CUPRAC assay were not statistically different according to the production method (organic/conventional) or botanical origin, with the exception of the sample of organically produced linden honey. No significant correlation was found between the results of the CUPRAC assay and the content of the phenolic compounds analyzed (Table S3). This indicates the ability of the non-phenolic constituents in honey to reduce Cu^{+2} ions, which should be interesting to examine in future work. In the literature data available to us, there is limited data on the antioxidant properties of honey obtained using the CUPRAC assays [95]. For example, the antioxidant activity of chestnut honey was registered from 11.00 to 97.07 mmol Trolox/100 g using this method [96].

3.5. Mineral Composition

The mineral content of nectar honey is generally low and ranges between 0.02% and 0.3% [97]. The mineral composition of honey is influenced by several factors, such as soil and climatic conditions, the chemical composition of the nectar (which varies according to the different botanical sources) and beekeeping techniques [29]. The most abundant mineral elements in the samples analyzed were potassium (2225.56 µg/g), phosphorus (923.92–795.40 µg/g) and calcium (15.70–240.60 µg/g), with the greatest differences between samples in calcium content (Table 4). In addition to these macroelements, significant content of other macroelements, such as magnesium, sodium and sulfur, was found in all examined samples (Table 4), which is consistent with literature data [29,98]. Potassium is the most abundant element in honey, accounting for one-third of the total mineral composition, which may be a consequence of its rapid secretion by the nectaries, and the potassium content can be more than ten times higher than the content of other macroelements in honey [98]. Less abundant are the elements iron (0.91–5.44 µg/g), manganese (0.09–6.60 µg/g) and zinc (0.22–3.75 µg/g), which belong to the group of microelements (Table 4). Of the toxic elements in the analyzed samples, the presence of boron (1.71–7.54 µg/g) and aluminum (0.99–4.31 µg/g) was detected in the highest concentration, while the presence of toxic elements such as lead and arsenic was not recorded, and lithium was practically in traces (0.008–0.017 µg/g; Table 4). The presence of toxic elements in honey is actually a consequence and indicator of environmental pollution [29]. Therefore, the mineral composition of honey is considered an indicator of environmental pollution [99].

As the mineral composition of honey is a direct result of its presence in the environment, published studies have also found wide variation in its presence in honey. For example, the magnesium content of honey was found to range from 2.18 to 563.72 mg/kg, the iron content from 0.41 to 224.00 mg/kg, or the zinc content from 0.05 to 17.30 mg/kg of honey [29,100].

Table 4. Content of mineral elements of tested honey samples [a].

Element	Organic Produced				Conventionally Produced				Pooled Std
	Linden	Acacia	Chestnut	Meadow	Linden	Acacia	Chestnut	Meadow	
macroelements (μg/g)									
Ca	169.46 [a]	15.70 [b]	78.48 [c]	25.49 [d]	147.25 [e]	15.29 [f]	240.60 [g]	121.36 [h]	0.02
K	1248.81 [a]	191.41 [b]	1305.66 [c]	1346.82 [d]	1281.12 [e]	183.91 [f]	2225.56 [g]	366.32 [h]	2.04
Mg	23.73 [a]	6.86 [b]	52.31 [c]	98.00 [d]	22.56 [a]	5.93 [e]	51.39 [c]	28.89 [f]	0.54
Na	13.28 [a]	12.99 [a]	46.82 [b]	13.13 [a]	12.57 [a]	12.84 [a]	16.93 [c]	14.28 [d]	0.54
P	807.18 [a]	895.24 [b]	853.64 [c]	923.92 [d]	840.57 [e]	795.40 [f]	922.59 [d]	874.92 [g]	2.02
S	27.58 [a]	15.65 [b]	56.37 [c]	95.36 [d]	25.34 [e]	17.61 [f]	40.31 [g]	33.07 [h]	0.82
microelements (μg/g)									
Co	0.08 [a]	n.d	n.d	0.02 [b]	n.d	0.01 [b]	n.d	n.d	0.008
Cr	0.05 [a]	0.05 [a]	0.06 [a]	0.10 [b]	0.06 [a]	0.03 [a]	0.10 [b]	0.05 [a]	0.02
Cu	0.11 [a]	0.10 [a]	0.41 [b]	0.84 [c]	0.09 [ad]	0.07 [d]	0.28 [be]	0.18 [e]	0.04
Fe	5.44 [a]	1.10 [b]	3.63 [c]	3.59 [c]	1.00 [bd]	0.91 [d]	1.52 [e]	0.91 [d]	0.09
Mn	2.22 [a]	0.30 [b]	6.60 [c]	5.42 [d]	0.67 [e]	0.09 [f]	3.48 [g]	0.20 [h]	0.05
Ni	0.05 [ab]	0.08 [b]	0.30 [c]	0.47 [d]	0.04 [a]	0.08 [ab]	0.12 [e]	0.17 [f]	0.04
Sr	0.59 [a]	0.03 [b]	0.19 [c]	0.04 [b]	0.11 [d]	0.03 [b]	0.22 [c]	0.05 [e]	0.03
Zn	1.19 [a]	0.58 [b]	0.74 [c]	3.75 [d]	0.38 [e]	0.22 [f]	1.23 [a]	1.86 [g]	0.04
toxic elements (μg/g)									
Al	1.39 [a]	1.19 [bc]	4.31 [d]	2.75 [e]	0.99 [b]	1.16 [c]	1.59 [a]	1.21 [bc]	0.05
As	n.d	n.d	n.d	n.d	n.d	n.d	n.d	n.d	-
B	2.24 [a]	2.43 [b]	1.71 [c]	3.81 [d]	2.90 [e]	3.09 [f]	2.27 [a]	7.54 [g]	0.04
Ba	0.27 [a]	0.01 [b]	0.23 [a]	0.05 [c]	0.05 [c]	0.01 [b]	1.11 [d]	0.02 [e]	0.02
Cd	n.d	n.d	n.d	0.01	n.d	n.d	n.d	n.d	0.002
Li	0.01 [a]	0.009 [a]	0.017 [b]	0.008 [a]	0.009 [a]	0.007 [c]	0.011 [b]	0.010 [b]	0.001
Pb	n.d	n.d	n.d	n.d	n.d	n.d	n.d	n.d	-
TMMEC (μg/g)	2299.77	1140.09	2405.21	2516.95	2331.76	1032.42	3504.33	1442.26	-
TTEC (μg/g)	3.91	3.64	6.27	6.63	3.95	4.27	4.98	7.57	-
TMEC (μg/g)	2303.68	2411.48	2411.477	2523.588	2335.71	1036.69	3509.31	1449.83	-

[a] Means in the same row with different letters are a significant difference according to the *t*-test ($p < 0.05$). Data are expressed as mean and pooled standard deviation (Pooled std) of three replicates. n.d.-not detected. TMMEC-total macro- and micoelements content; TTEC—total toxic elements content; TMEC—total mineral content.

3.6. Raman Spectroscopy Analysis with PCA Analysis

The recorded Raman spectra of different honey samples were presented in the fingerprint region, i.e., between 300 and 1500 cm^{-1} (Figure 2A). The characteristic band in all honey samples is identified at 353 cm^{-1} and can be assigned to the δ(C-C-C) ring vibration of carbon hydrates [101]. The bands recorded at 422 cm^{-1}, 519 cm^{-1} and 628 cm^{-1} [102] could be assigned to δ(C-C-O) [103], δ(C-C-C) carbohydrates [102,103], and ring deformation, respectively [102,104]. The band at 709 cm^{-1} contributes to ν(C-O), ν(C-C-C) and δ(O-C-O) [102]. The bands at 821 cm^{-1}, 867 cm^{-1} and 920 cm^{-1} contribute to ν(C-O-H) [102,105], δ(C-H) [102,106], and δ(C-O-H) [102,107], respectively. Intensive bands were also identified at 1061 cm^{-1} and 1124 cm^{-1}. The first may be attributed to ν(C-C), ν(C-O) and δ(C-O-H) carbohydrates [106]. The second band is most likely linked to ν(C-O) and δ(C-O-H) chemical bonds [102,103].

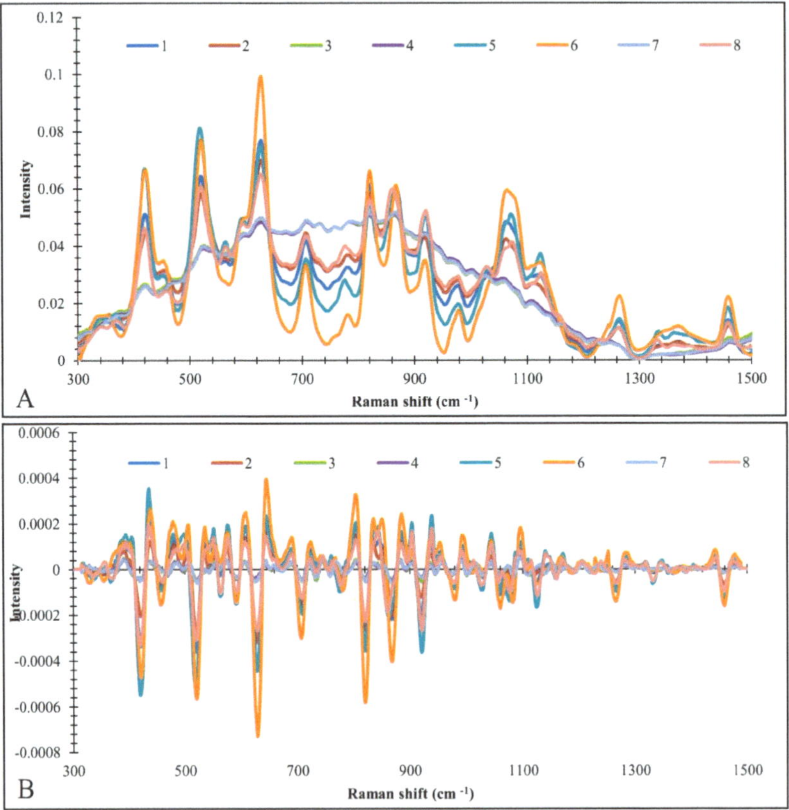

Figure 2. Raw spectra (**A**) and pre-processed spectra (baseline correction + normalization + 2nd order derivative and smoothing (**B**). Legend: 1—linden organic; 2—accacia organic; 3—chestnut organic; 4—meadow organic; 5—linden conventional; 6—accacia conventional; 7—chestnut conventional; 8-meadow conventional.

The results of the PCA are presented by score and loading plots (Figure 3). The score plot shows a clear tendency to group the different honey samples (Figure 3A). According to the PC1 axis, samples 5 and 6 and samples 3, 4 and 7 were grouped jointly. The results of the loading plot indicate that samples 5 and 6 have been separated as one distinct class due to the negative results at ~422, ~519 and ~628 cm^{-1}. Samples 3, 4 and 7 were grouped in one separate cluster due to the positive loading values of PC1 at ~820 cm^{-1}. On the other hand, samples 1 and 5, in addition to samples 2 and 6, were grouped together due to the PC2 values. The results of the loading plot indicate that strong bands at ~628, ~821, and ~867 cm^{-1} were responsible for the grouping of samples 2 and 6, whereas bands at ~422, ~920 cm^{-1} determined the separation of samples 1 and 5 into one distinct class (Figure 2B).

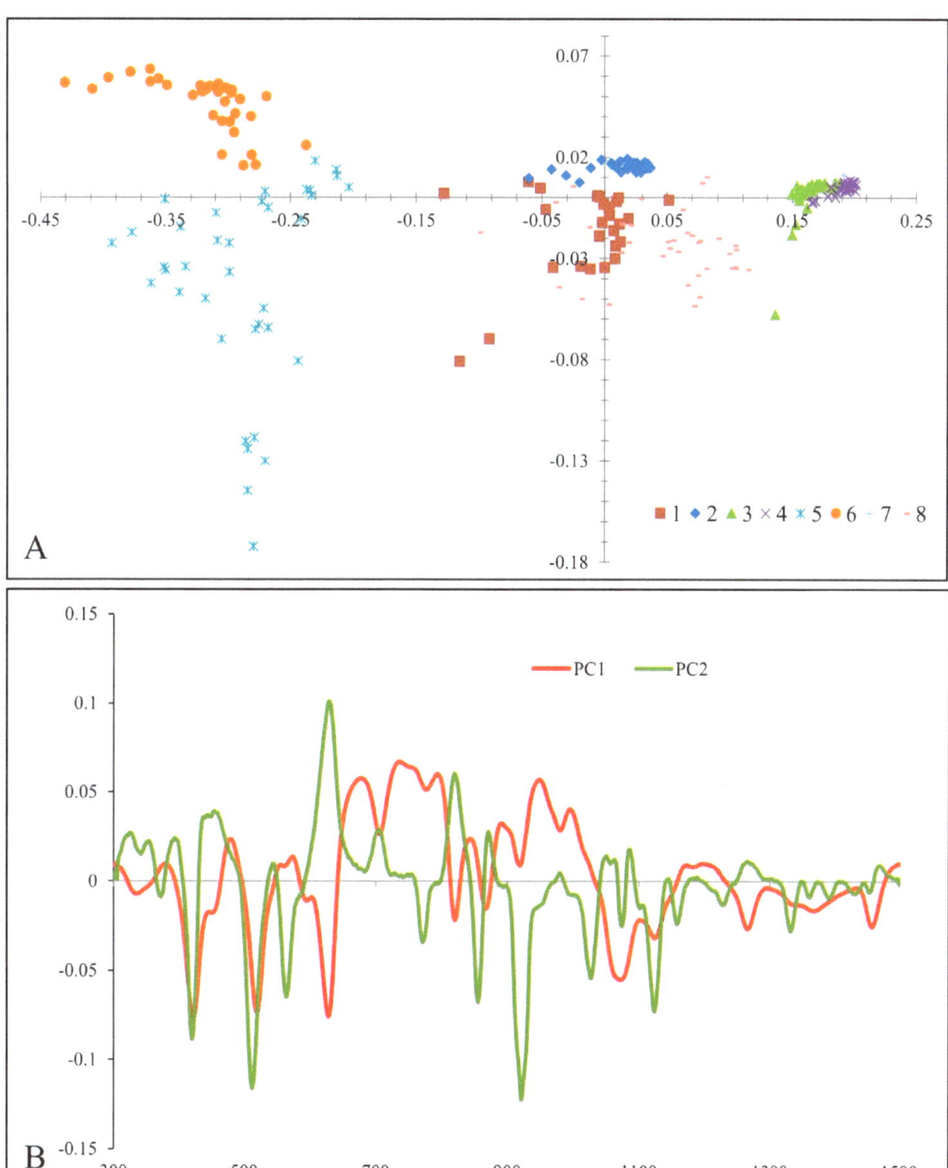

Figure 3. PCA score plot (**A**) and Loading plot (**B**). Legend: 1—linden organic; 2—accacia organic; 3—chestnut organic; 4—meadow organic; 5—linden conventional; 6—accacia conventional; 7—chestnut conventional; 8—meadow conventional.

The results of the classification of different honey samples are shown in Tables S4 and S5. In both models tested, the 5 PCs were used, explaining 99% of the total variability. In the first case, using the model based on 8 classes (corresponding with 8 studied honey samples), the accuracy was between 83.33–100.00% (96.88% in total) (Table S4). In the second case, using 4 classes based on the botanical origin of the studied honey samples, the model accuracy ranged between 84.72 and 98.61% (92.36% in total). The graphical representation of the discrimination results is presented in Figures S1 and S2.

Madgas et al. [108] used Raman spectroscopy and chemometrics to classify different honey types. Based on Soft Independent Modeling of Class Analogy (SIMCA), their results showed high percentage accuracy for acacia (100%), chestnut (100%) and linden (83%). Oroian and Ropciuc [101] applied Raman spectroscopy and linear discriminant analysis (LDA) to determine the botanical origin of different honey samples. Accordingly, honeydew samples were correctly classified in 95% of the studied samples, while in the case of Acacia honey, the accuracy was 90%.

To the best of our knowledge, there is no work based on the application of Raman spectroscopy and chemometrics in distinguishing the honey samples originating from different production sites, which were simultaneously distinct production systems (organic and conventional). Our results indicate that Raman spectroscopy associated with the appropriate chemometric modeling has successfully classified different honey samples. However, fine differences in spectra of the same honey types obtained from the different production systems are most likely the consequence of a specific chemical composition of a sample reflecting peculiarities of different honey-bee collecting sites -locations (bee pastures), rather than production systems (organic versus conventional) since organic (certified) and conventional honey production sites have to be sufficiently distant. Finally, there was no typical Raman band specifically corresponding to either organic or conventional honey samples. However, Raman spectroscopy showed very high validity for the classification of different honey samples based on their botanical origin.

3.7. Sensory Analysis

Sensory analysis of honey is a fast and practical way to obtain information about the quality of honey and is often used to determine the price of honey [109]. It can also detect undesirable characteristics that are not reached by routine analysis, such as metallic taste, fermentation, smoky odor, or the presence of impurities [17]. Honey is characterized by specific sensory properties due to the large number of components that come from both the nectar and the bees themselves. When comparing organic and conventional honey samples using the hedonic rating scale, higher scores were obtained for the overall sensory acceptability of samples from organic beekeeping (except for linden honey, where the scores are very close to each other; Table 1). Organic acacia honey scored the highest for overall acceptability (7.7), while conventional chestnut honey scored the lowest (4.1). The reason for the low overall acceptability rating of conventional chestnut honey could be due to experts' assessment of spiciness and bitterness (Figure 4). Six out of ten expert evaluators stated that the smell of the organic acacia honey was not present, while 8 out of 10 evaluators declared that the conventional chestnut honey had a strong smell (Figure 4). According to Bogdanov [13], over 600 aromatic compounds were detected in different types of honey. Most of the volatile compounds come from the flower of the plant and certain monofloral honey. Certain volatiles are found exclusively in certain types of honey and are used to accurately test the botanical origin of honey [26]. The evaluators stated that chestnut and meadow honey were the darkest in color, corresponding to dark and light amber shades of the Pfund scale (Figure 4; Table 1). None of the samples tested were found to have undesirable characteristics, such as fermentation, metallic taste, smoky odors or the presence of impurities (Figure 4). The sensory characteristics of the tested honey samples clearly depended on the botanical origin but not on organic or conventional beekeeping.

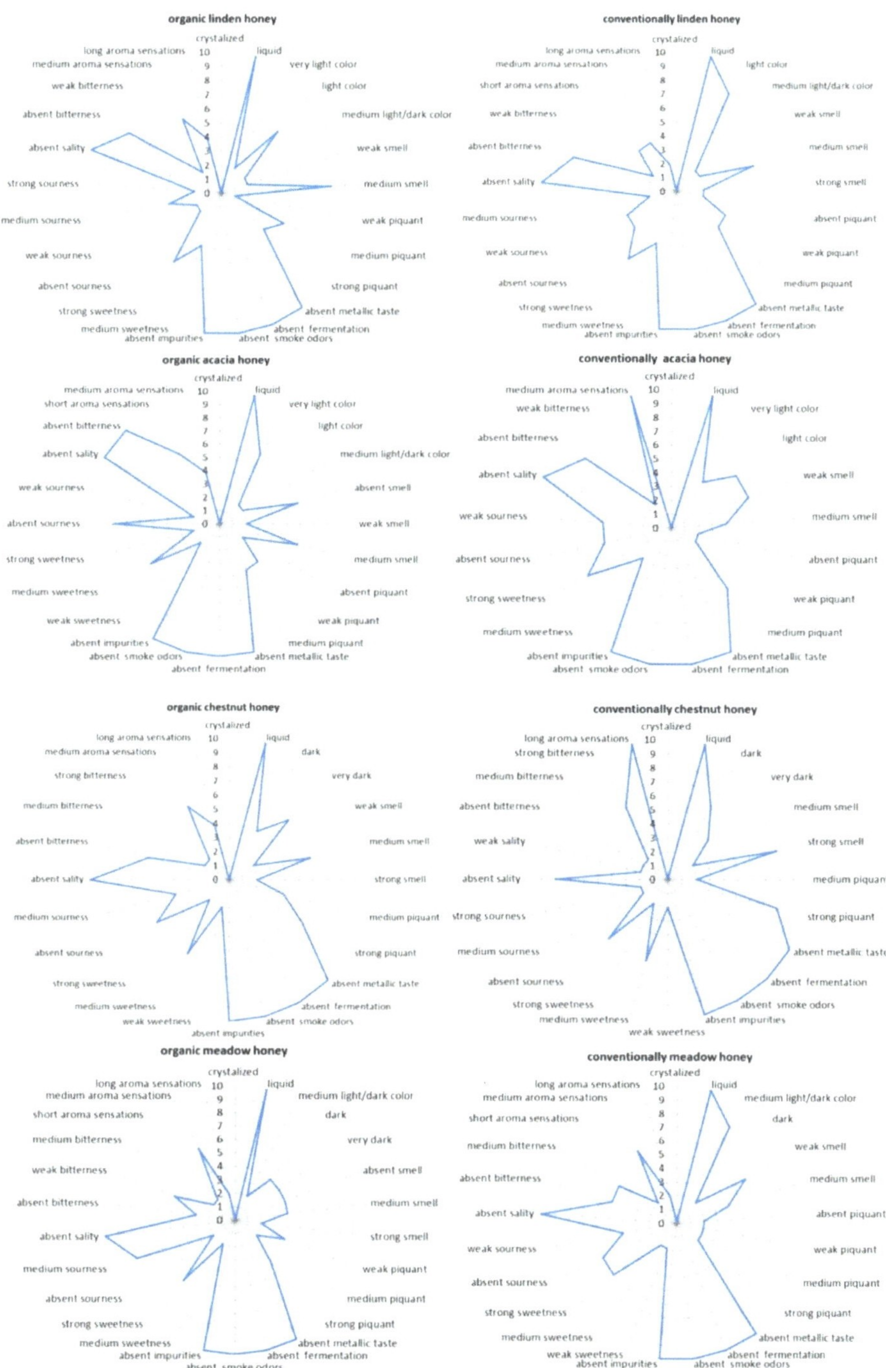

Figure 4. The result of the sensory evaluation of the tested honey samples by trained evaluators (n = 2 × 10).

4. Conclusions

Although statistically significant differences in physicochemical parameters (specific optical rotation, electrical conductivity, moisture, diastase activity, free acidity, pH) were found between all samples of organic and conventional honey of the corresponding botanical species, no general trend in the parameters depending on the beekeeping method can be established. These slight differences within the same botanical origin may be due to different geographical origins, although all samples are from the Balkans. The mineral composition, total phenolic and total flavonoid contents, profile of phenolic compounds and antioxidant properties significantly depended on the botanical origin of the honey and not on the beekeeping method. Organic honey samples were only richer in hydroxycinnamic acid derivatives. In the phenolic profile of the analyzed samples, 38 phenolic compounds and 4 phenylamides were identified, with the largest proportion contributed in most samples coming from propolis-derived flavonoid aglycones (pinocembrin, chrysin, pinobanksin and galangin). Raman spectroscopy did not show the differentiation of honey according to the beekeeping method (organic/conventional) but showed the botanical origin. No clear differences were observed between the sensory properties of honey samples from organic and conventional beekeeping. The PCA analysis did not reveal any general differences between organic and conventional honey samples. The general conclusion is that the physiochemical, nutritional and sensory characteristics do not depend significantly on the method of honey production (organic or conventional beekeeping) but much more on the botanical origin. However, as legal regulations prescribe the absence of pesticides and other anti-nutritive components in organic honey, in the future, differences in the content of these components between samples obtained from organic and conventional beekeeping could be examined. In this way, a complete "picture" of both the quality and the safety of the tested samples would be obtained. Also, in future studies, a larger number of samples should be considered in order to confirm the obtained statistical models.

Supplementary Materials: The following supporting information can be downloaded at https://www.mdpi.com/article/10.3390/foods13223573/s1, Table S1. Botanical and geographical origin of examined honey samples; Table S2. Equation parameters of used phenolic standards for quantification; Table S3. Correlation coefficients (*r*) between quality parameters of tested honey samples; Table S4. Classification results of QDA model with 8 classes; Table S5. Classification results of QDA model with 4 classes (botanical origin); Figure S1. QDA discrimination plots; Figure S2. QDA discrimination plots.

Author Contributions: Conceptualization, S.P.S., M.B.P. and A.Ž.K.; Methodology, S.P.S.; Validation, S.P.S., D.D.M., N.S., S.K. and A.Ž.K.; Formal analysis, S.P.S., D.D.M., N.S., N.M.N., S.K., B.D., Z.D.-S. and A.Ž.K.; Investigation, S.P.S.; Resources, N.S. and N.M.N.; Writing—original draft, S.P.S., D.D.M., M.B.P. and S.K.; Writing—review & editing, S.P.S. and M.B.P.; Visualization, S.P.S. and M.B.P.; Supervision, S.P.S.; Project administration, S.P.S., M.B.P. and B.D.; Funding acquisition, S.P.S. All authors have read and agreed to the published version of the manuscript.

Funding: This work was supported by the Ministry of Science, Technological Development and Innovation of the Republic of Serbia (Grant No. 451-03-65/2024-03/200116 and Grant No. 451-03-66/2024-03/200026), and by the Science Fund of Republic of Serbia (#Grant No. 7744714, FUNPRO and #GRANT No. 7750160, R-SPECT).

Institutional Review Board Statement: Not applicable.

Informed Consent Statement: Not applicable.

Data Availability Statement: The original contributions presented in the study are included in the article/Supplementary Material, further inquiries can be directed to the corresponding author.

Acknowledgments: The authors are grateful to the beekeepers who provided honey samples for analysis in this study.

Conflicts of Interest: The authors declare that they have no known competing financial interests or personal relationships that could have appeared to influence the work reported in this paper.

References

1. Codex Alimentarius. Internation Food Standards, Standard for Honey CXS 12-19811, Adopted in 1981. Revised in 1987, 2001. Amended in 2019, 2022. Available online: https://www.fao.org/fao-who-codexalimentarius/codex-texts/list-standards/en/ (accessed on 28 July 2024).
2. Cartwright, M. Medicine in the Ancient World. 2019. Available online: https://www.worldhistory.org/collection/59/medicine-in-the-ancient-world/ (accessed on 17 July 2024).
3. Biswajit, P.; Surya, N.P. Contamination of honey: A human health perspective. In *Health Risks of Food Additives-Recent Developments and Trends in Food Sector*; IntechOpen: London, UK, 2023. [CrossRef]
4. Bogdanov, S.; Jurendić, T.; Sieber, R.; Gallmann, P. Honey for nutrition and health, A review. *Am. J. Coll. Nutr.* **2008**, *27*, 677–689. [CrossRef] [PubMed]
5. Official Gazette of the RS. Rulebook on Control and Certification in Organic Production and Organic Production Methods No: 95. 2021. Available online: https://www.pravno-informacioni-sistem.rs/SlGlasnikPortal/eli/rep/sgrs/ministarstva/pravilnik/2020/95/1/reg/ (accessed on 1 October 2024). (In Serbian).
6. Regulation EC (European Commission Regulation). Regulation No 396/2005 of the European Parliament and of the Council on Maximum Residue Levels of Pesticides in or on Food and Feed of Plant and Animal Origin and Amending Council Directive 91/414/EEC. 2005. Available online: https://eur-lex.europa.eu/legal-content/EN/ALL/?uri=celex:32005R0396 (accessed on 30 July 2024).
7. Regulation EC (European Union Regulation). Commission regulation No 37/2010 on pharmacologically active substances and their classification regarding maximum residue limits in foodstuffs of animal origin. *Off. J. Eur. Union* **2010**, L15/1–L15/72. Available online: https://eur-lex.europa.eu/LexUriServ/LexUriServ.do?uri=OJ:L:2010:015:0001:0072:en:PDF (accessed on 30 July 2024).
8. Castroa, E.; Quicaz, M.; Mojicab, A.; Zuluaga-Domínguez, C. Bioactive and physicochemical profile of honey collected from Colombian organic and conventional coffee growing areas. *J. Apic. Res.* **2023**, *62*, 518–529. [CrossRef]
9. Halagarda, M.; Groth, S.; Popek, S.; Rohn, S.; Pedan, V. Antioxidant activity and phenolic profile of selected organic and conventional honeys from Poland. *Antioxidants* **2020**, *9*, 44. [CrossRef]
10. Polak-Śliwińska, M.; Tańska, M. Conventional and organic honeys as a source of water-and ethanol-soluble molecules with nutritional and antioxidant characteristics. *Molecules* **2021**, *26*, 3746. [CrossRef]
11. USS. United States Standards for Grades of Extracted Honey. 1985. Available online: https://www.ams.usda.gov/sites/default/files/media/Extracted_Honey_Standard[1].pdf (accessed on 19 July 2024).
12. IHC (International Honey Commission). Harmonised Methods of the International Honey Commission. 2009. Available online: https://www.ihc-platform.net/ihcmethods2009.pdf (accessed on 29 July 2024).
13. Bogdanov, S. Honey Composition. In *The Honey Book, Chapter 5. Bee Product Science*; Hastings House: London, UK, 2011; pp. 1–10. Available online: https://www.academia.edu/5616849/Composition_of_honey (accessed on 22 July 2024).
14. Kostić, A.Ž.; Milinčić, D.D.; Špirović Trifunović, B.; Nedić, N.; Gašić, U.M.; Tešić, Ž.L.; Stanojević, S.P.; Pešić, M.B. Monofloral corn poppy bee-collected pollen-a detailed insight into its phytochemical composition and antioxidant properties. *Antioxidants* **2023**, *12*, 1424. [CrossRef]
15. Kostić, A.Ž.; Milinčić, D.D.; Nedić, N.; Gašić, U.M.; Špirović Trifunović, B.; Vojt, D.; Tešić, Ž.L.; Pešić, M.B. Phytochemical profile and antioxidant properties of bee-collected artichoke (*Cynara scolymus*) Pollen. *Antioxidants* **2021**, *10*, 1091. [CrossRef]
16. Kostić, Ž.A.; Pešić, B.M.; Mosić, D.M.; Dojčinović, P.B.; Natić, N.M.; Trifković, Đ.J. Mineral content of some bee-collected pollen from Serbia. *Arch. Ind. Hyg. Toxicol.* **2015**, *66*, 251–258. [CrossRef]
17. Marcazzan, G.L.; Mucignat-Caretta, C.; Marchese, C.M.; Pianad, M.L. A review of methods for honey sensory analysis. *J. Apic. Res.* **2018**, *57*, 175–187. [CrossRef]
18. Stanojevic, P.S.; Barać, B.M.; Pešić, B.M.; Vucelic-Radovic, V.B. Protein composition and textural properties of inulin-enriched tofu produced by hydrothermal process. *LWT Food Sci. Technol.* **2020**, *126*, 109309. [CrossRef]
19. SUBg (Senate of the University of Belgrade). The Code of Professional Ethics of the University of Belgrade. *Off. Gaz. Repub. Serb.* **2016**, *189*, 16.
20. Gangwar, S.K.; Gebremariam, H.; Ebrahim, A.; Tajebe, S. Characteristics of honey produced by different plant species in Ethiopia. *Adv. Biores.* **2010**, *1*, 101–105. Available online: https://www.researchgate.net/publication/308095735 (accessed on 1 July 2024).
21. Sanchez, M.P.; Huidobro, J.F.; Mato, I.; Muniategui, S.; Sancho, T. Correlation between proline content of honeys and botanical origin. *Dtsch. Lebensm. Rundsch.* **2001**, *97*, 171175.
22. Codex Alimentarius Commission. Revised Codex Standard for Honey, Codex Standard 12-1981, Rev. 2. 2001. Available online: https://www.fao.org/3/w0076e/w0076e30.htm (accessed on 11 July 2024).
23. Kulinčević, J. *Beekeeping*; National Library of Serbia: Belgrade, Serbia, 2016; pp. 141–154. (In Serbian)
24. Bogdanov, S.; Ruoff, K.; Persano Oddo, L. Physico-chemical methods for the characterisation of unifloral honeys: A review. *Apidologie* **2004**, *35*, S4–S17. [CrossRef]
25. Persano Oddo, L.; Piazza, M.G.; Sabatini, A.G.; Accorti, M. Characterization of unifloral honeys. *Apidologie* **1995**, *26*, 453–465. [CrossRef]

26. Bogdanov, S.; Lullmann, C.; Martin, P.; Ohe, W.; Russmann, H.; Vorwohl, G.; Oddo, L.; Sabatini, A.; Marcazzan, G.; Piro, R.; et al. Honey quality and international regulatory standards, review by the International honey commission. *Bee World* **2015**, *80*, 61–69. [CrossRef]

27. Persano Oddo, L.; Piro, R. Main European unifloral honeys: Descriptive sheets. *Apidologie* **2004**, *35*, S38–S81. [CrossRef]

28. Bogdanov, S.; Haldimann, M.; Luginbuhl, W.; Gallmann, P. Minerals in honey: Environmental, geographical and botanical aspects. *J. Apic. Res.* **2007**, *46*, 269–275. [CrossRef]

29. Solayman, M.; Islam, M.A.; Paul, S.; Ali, Y.; Khalil, M.I.; Alam, N.; Gan, S.H. Physicochemical properties, minerals, trace elements, and heavy metals in honey of different origins, A Comprehensive Review. *Compr. Rev. Food Sci. Food Saf.* **2016**, *15*, 219–233. [CrossRef]

30. Regulation EC (European Commission Regulation). Health and Consumer Protection Directorate-General. Opinion of the Scientific Committee on Veterinary Measures Relating to Public Health on Honey and Microbiological Hazards. 2002. Available online: https://food.ec.europa.eu/document/download/7488a863-4179-444a-8eff-72b137ae0240_en?filename=sci-com_scv_out53_en.pdf (accessed on 19 July 2024).

31. Živkov Baloš, M.; Popov, N.; Vidaković, S.; Ljubojević Pelić, D.; Pelić, M.; Mihaljev, Ž.; Jakšić, S. Selectrical conductivity and acidity of honey. *Arch. Vet. Med.* **2018**, *11*, 91–101. [CrossRef]

32. Official Gazette of the RS. Rulebook on the Quality of Honey and Other Bee Products: 101/2015-114. 2015. Available online: https://pravno-informacioni-sistem.rs/eli/rep/sgrs/ministarstva/pravilnik/2015/101/2 (accessed on 25 July 2024). (In Serbian).

33. Codex Alimentarius. Draft Revised for Honey at Step 6 of the Codex Procedure. CX 5/10.2, CL1998/12-S 1998. Available online: https://www.fao.org/3/X4616E/x4616e07.htm (accessed on 15 July 2024).

34. Regulation EC (European Commission Regulation). Council Directive 2001/110/EC of 20 December 2001 Relating to Honey. 2021. Available online: https://eur-lex.europa.eu/LexUriServ/LexUriServ.do?uri=OJ:L:2002:010:0047:0052:EN:PDF (accessed on 2 July 2024).

35. Chirifie, J.; Zamora, M.C.; Motto, A. The correlation between water activity and % moisture in honey: Fundamental aspects and application to Argentine honeys. *J. Food Eng.* **2006**, *72*, 287–292. [CrossRef]

36. Escuredo, O.; Míguez, M.; Fernández-González, M.; Seijo, M.C. Nutritional value and antioxidant activity of honeys produced in a European Atlantic area. *Food Chem.* **2013**, *138*, 851–856. [CrossRef] [PubMed]

37. Machado, M.A.; Tomas, A.; Russo-Almeida, P.; Duarte, A.; Antunes, M.; Vilas-Boas, M.; Graça, M.M.; Figueiredo, A.C. Quality assessment of Portuguese monofloral honeys. Physicochemical parameters as tools in botanical source differentiation. *Food Res. Int.* **2022**, *157*, 111362. [CrossRef] [PubMed]

38. Ojeda, G.; Sulbaran, B.; Ferrer, A.; Rodriguez, B. Characterization of honey produced in Venezuela. *Food Chem.* **2004**, *84*, 499–502. [CrossRef]

39. Persano Oddo, L.; Piazza, M.G.; Pulcini, P. Invertase activity in honey. *Apidologie* **1999**, *30*, 57–65. [CrossRef]

40. White, J.W., Jr. Composition of honey. In *Honey: A Comprehensive Survey*; Crane, E., Ed.; Heinemann: London, UK, 1975; pp. 180–194.

41. Persano Oddo, L.; Baidi, E.; Accorti, M. Diastatic activity in some unifloral honeys. *Apidologie* **1990**, *21*, 17–24. [CrossRef]

42. Missio da Silva, P.; Gonzaga, L.V.; Biluca, F.C.; Schulz, M.; Vitali, L.; Micke, G.A.; Oliveira Costa, A.C.; Fett, R. Stability of Brazilian *Apis mellifera* L. honey during prolonged storage: Physicochemical parameters and bioactive compounds. *LWT Food Sci. Technol.* **2020**, *129*, 109521. [CrossRef]

43. Missio da Silva, P.M.; Gauche, C.; Gonzaga, L.V.; Costa, A.C.O.; Fett, R. Honey: Chemical composition, stability and authenticity. *Food Chem.* **2016**, *196*, 309–323. [CrossRef]

44. Alqarni, A.S.; Owayss, A.A.; Mahmoud, A.A. Physicochemical characteristics, total phenols and pigments of national and international honeys in Saudi Arabia. *Arab. J. Chem.* **2016**, *9*, 114–120. [CrossRef]

45. Mohammed, S.A.; Babiker, E.E. Protein structure, physicochemical properties and mineral composition of *Apis mellifera* Honey samples of different floral origin. *Aust. J. Basic Appl. Sci.* **2009**, *3*, 2477–2483.

46. Halliwell, B.; Gutteridge, J.M.C. *Free Radicals in Biology and Medicine*, 5th ed.; Oxford University Press: New York, NY, USA, 2015. [CrossRef]

47. Šarić, G.; Matković, D.; Hruškar, M.; Vahčić, N. Characterisation and classification of Croatian honey by physicochemical parameters. *Food Technol. Biotechnol.* **2008**, *46*, 355–367.

48. Perez-Arquillué, C.; Conchello, P.; Arino, A.; Juan, T.; Herrera, A. Quality evaluation of Spanish rosemary (*Rosmarinus officinalis*) honey. *Food Chem.* **1994**, *51*, 207–210. [CrossRef]

49. Rysha, A.; Kastrati, G.; Biber, L.; Sadiku, V.; Rysha, A.; Zogaj, F.; Kabashi-Kastrati, E. Evaluating the physicochemical properties of some Kosovo's and imported honey samples. *Appl. Sci.* **2022**, *12*, 629. [CrossRef]

50. Fell, R.D. The color grading of honey. *Am. Bee J.* **1978**, *118*, 782–789.

51. Bertoncelj, J.; Doberšek, U.; Jamnik, M. Evaluationof the phenolic content, antioxidant activity and colour of Slovenian honey. *Food Chem.* **2007**, *105*, 822–828. [CrossRef]

52. Ramalhosa, E.E.; Gomes, T.T.; Pereira, A.P.; Dias, T.T.; Estevinho, L.M. Mead production tradition versus modernity. *Adv. Food Nutr. Res.* **2011**, *63*, 101–118. [CrossRef]

53. Brudzynski, K.; Kim, L. Storage-induced chemical changes in active components of honey de-regulate its antibacterial activity. *Food Chem.* **2011**, *126*, 1155–1163. [CrossRef]

54. Moniruzzaman, M.; Khalil, M.I.; Sulaiman, S.A.; Gan, S.H. Physicochemical and antioxidant properties of Malaysian honeys produced by *Apis cerana*, *Apis dorsata* and *Apis mellifera*. *BMC Complement. Altern. Med.* **2013**, *13*, 1–12. Available online: http://www.biomedcentral.com/1472-6882/13/43 (accessed on 10 July 2024). [CrossRef]
55. Amiot, M.J.; Aubert, S.; Gonnet, M.; Tacchini, M. Phenolic composition of honeys: Preliminary study on identification and group quantification. *Apidologie* **1989**, *20*, 115–125. [CrossRef]
56. Gómez-Caravaca, A.M.; G ómez-Romero, M.; Arráez-Román, D.; Segura-Carretero, A.; Fernández-Gutiérrez, A. Advances in the analysis of phenolic compounds in products derived from bees. *J. Pharm. Biomed. Anal.* **2006**, *41*, 1220–1234. [CrossRef]
57. AL-Mamary, M.; AL-Meeri, A.; AL-Habori, M. Antioxidant activities and total phenolics of different types of honey. *Nutr. Res.* **2002**, *22*, 1041–1047. [CrossRef]
58. Gheldof, N.; Engeseth, N.J. Antioxidant capacity of honeys from various floral sources based on the determination of oxygen radical absorbance capacity and inhibition of in vitro lipoprotein oxidation in human serum samples. *J. Agric. Food Chem.* **2002**, *50*, 3050–3055. [CrossRef] [PubMed]
59. Baek, Y.; Kim, Y.J.; Baik, M.-Y.; Kim, D.-O. Total phenolic contents and antioxidant activities of Korean domestic honey from different floral sources. *Food Sci. Biotechnol.* **2015**, *24*, 1453–1457. Available online: https://www.researchgate.net/publication/28 2303853 (accessed on 12 July 2024). [CrossRef]
60. Moniruzzaman, M.; Sulaiman, S.A.; Mohd AS, A.; Hua Gan, S. Two-year variations of phenolics, flavonoids and antioxidant contents in acacia honey. *Molecules* **2013**, *18*, 14694–14710. [CrossRef] [PubMed]
61. Azad MA, K.; Tong, Q.; Al-Faruq, A. Antioxidant activity of some monofloral honeys: Different contributions of the raw honey and phenolic extract. *Int. Res. J. Biol. Sci.* **2016**, *5*, 45–50.
62. Akgün, N.; Çelik, Ö.F.; Kelebekli, L. Physicochemical properties, total phenolic content, and antioxidant activity of chestnut, rhododendron, acacia and multifloral honey. *J. Food Meas. Charact.* **2021**, *15*, 3501–3508. [CrossRef]
63. Karabagias, I.K.; Maia, M.; Karabagias, V.K.; Gatzias, I.; Badeka, A.V. Quality and origin characterisation of Portuguese, Greek, Oceanian, and Asian honey, based on poly-parametric analysis hand in hand with dimension reduction and classification techniques. *Eur. Food Res. Technol.* **2020**, *246*, 987–1006. [CrossRef]
64. Marić, A.; Jovanov, P.; Sakač, M.; Novaković, A.; Hadnadev, M.; Pezo, L.; Mandić, A.; Milićević, N.; Đurović, A.; Gadžurić, S.A. Comprehensive study of parameters correlated with honey health benefits. *RSC Adv.* **2021**, *11*, 12434–12441. [CrossRef]
65. Gorjanović, S.; Suarez, J.; Novaković, M.; Pastor, F.; Pezo, L.; Battino, M.; Sužnjević, D. Comparative analysis of antioxidant activity of honey of different floral sources using recently developed polarographic and various spectrophotometric assays. *J. Food Compos. Anal.* **2013**, *30*, 13–18. [CrossRef]
66. Ferreres, F.; Ortiz, A.; Silva, C.; García-Viguera, C.; Tomás-Barberán, F.A.; Tomás-Lorente, F. Flavonoids of "La Alcarria" honey: A study of their botanical origin. *Z. Lebensm. Unters. Forsch.* **1992**, *194*, 139–143. [CrossRef]
67. Dong, R.; Zheng, Y.; Xu, B. Phenolic profiles and antioxidant capacities of chinese unifloral honeys from different botanical and geographical sources. *Food Bioprocess Technol.* **2013**, *6*, 762–770. [CrossRef]
68. Živković, J.; Sunarić, S.; Stanković, N.; Mihajilov-Krstev, T.; Spasić, A. Total phenolic and flavonoid contents, antioxidant and antibacterial activities of selected honeys against human pathogenic bacteria. *Acta Pol. Pharm. Drug Res.* **2019**, *76*, 671–681. [CrossRef] [PubMed]
69. Becerril-Sánchez, A.L.; Quintero-Salazar, B.; Dublán-García, O.; Escalona-Buendía, H.B. Phenolic compounds in honey and their relationship with antioxidant activity, botanical origin, and color. *Antioxidants* **2021**, *10*, 1700. [CrossRef]
70. Saral, Ö. An investigation into chestnut honeys from artvin province in Turkiye: Their physicochemical properties, phenolic profiles and antioxidant activities. *Chem. Biodivers.* **2023**, *20*, e202201162. [CrossRef]
71. Contardi, M.; Lenzuni, M.; Fiorentini, F.; Summa, M.; Bertorelli, R.; Suarato, G.; Athanassiou, A. Hydroxycinnamic acids and derivatives formulations for skin damages and disorders: A review. *Pharmaceutics* **2021**, *13*, 999. [CrossRef]
72. Ruwizhi, N.; Aderibigbe, B.A. Cinnamic acid derivatives and their biological efficacy. *Int. J. Mol. Sci.* **2020**, *21*, 5712. [CrossRef]
73. Gardana, C.; Scaglianti, M.; Pietta, P.; Simonetti, P. Analysis of the polyphenolic fraction of propolis from different sources by liquid chromatography-tandem mass spectrometry. *J. Pharm. Biomed. Anal.* **2007**, *45*, 390–399. [CrossRef]
74. Bertoncelj, J.; Polak, T.; Kropf, U.; Korošec, M.; Golob, T. LC-DAD-ESI/MS analysis of flavonoids and abscisic acid with chemometric approach for the classification of Slovenian honey. *Food Chem.* **2011**, *127*, 296–302. [CrossRef]
75. Gašić, U.; Kečkeš, S.; Dabić, D.; Trifković, J.; Milojković-Opsenica, D.; Natić, M.; Tešić, Z. Phenolic profile and antioxidant activity of Serbian polyfloral honeys. *Food Chem.* **2014**, *145*, 599–607. [CrossRef]
76. Kečkeš, S.; Gašić, U.; Veličković, T.Ć.; Milojković-Opsenica, D.; Natić, M.; Tešić, Ž. The determination of phenolic profiles of Serbian unifloral honeys using ultra-high-performance liquid chromatography/high resolution accurate mass spectrometry. *Food Chem.* **2013**, *138*, 32–40. [CrossRef]
77. Tomás-Barberán, F.; Martos, I.; Ferreres, F.; Radovic, B.; Anklam, E. HPLC flavonoid profile as markers for the botanical origin of European honey. *J. Sci. Food Agric.* **2001**, *81*, 485–496. [CrossRef]
78. Medana, C.; Carbone, F.; Aigotti, R.; Appendino, G.; Baiocchi, C. Selective analysis of phenolic compounds in propolis by HPLC-MS/MS. *Phytochem. Anal.* **2008**, *19*, 32–39. [CrossRef]
79. Pellati, F.; Orlandini, G.; Pinetti, D.; Benvenuti, S. HPLC-DAD and HPLC-ESI-MS/MS methods for metabolite profiling of propolis extracts. *J. Pharm. Biomed. Anal.* **2011**, *55*, 934–948. [CrossRef]

80. Ristivojević, P.; Trifković, J.; Gašić, U.; Andrić, F.; Nedić, N.; Tešić, Ž.; Milojković-Opsenica, D. Ultrahigh-performance liquid chromatography and mass spectrometry (UHPLC-LTQ/Orbitrap/MS/MS) study of phenolic profile of Serbian poplar type propolis. *Phytochem. Anal* **2015**, *26*, 127–136. [CrossRef]

81. Glavnik, V.; Bensa, M.; Vovk, I.; Guzelmeric, E. High-performance thin-layer chromatography–multi-stage mass spectrometry methods for analyses of bee pollen botanically originating from sweet chestnut (*Castanea sativa* Mill.). *JPC J. Planar Chromatogr. Mod. TLC* **2023**, *36*, 471–482. [CrossRef]

82. Rodríguez-Flores, M.S.; Escuredo, O.; Seijo, M.C.; Rojo, S.; Vilas-Boas, M.; Falcão, S.I. Phenolic profile of castanea bee pollen from the Northwest of the Iberian Peninsula. *Separations* **2023**, *10*, 270. [CrossRef]

83. Truchado, P.; Ferreres, F.; Bortolotti, L.; Sabatini, A.G.; Tomás-Barberán, F.A. Nectar Flavonol rhamnosides are floral markers of acacia (Robinia pseudacacia) honey. *J. Agric. Food. Chem.* **2008**, *56*, 8815–8824. [CrossRef]

84. Ferreira, I.C.F.R.; Aires, E.; Barreira, J.C.M.; Estevinho, L.M. Antioxidant activity of Portuguese honey samples: Different contributions of the entire honey and phenolic extract. *Food Chem.* **2009**, *144*, 1438–1443. [CrossRef]

85. Savatović, S.; Dimitrijević, D.; Đilas, S.; Čanadanović-Bruneta, J.; Ćetković, G.; Tumbas, V.; Štajner, D. Antioxidant activity of three different serbian floral honeys. *Acta Period. Technol.* **2011**, *42*, 145–155. [CrossRef]

86. Blasa, M.; Candiracci, M.; Accorsi, A.; Piacentini, M.P.; Albertini, M.C.; Piatti, E. Raw *Millefiori* honey is packed full of antioxidants. *Food Chem.* **2006**, *97*, 217–222. [CrossRef]

87. Buratti, S.; Benedetti, S.; Cosio, M.S. Evaluation of the antioxidant power of ho-ney, propolis and royal jelly by amperometric flow injection analysis. *Talanta* **2007**, *71*, 1387–1392. [CrossRef] [PubMed]

88. Beretta, G.; Granata, P.; Ferrero, M.; Orioli, M.; Facino, R.M. Standardization of antioxidant properties of honey by a combination of spectrophotometric/fluorimetric assays and chemometrics. *Anal. Chim. Acta* **2005**, *533*, 185–191. [CrossRef]

89. Larsen, P.; Ahmed, M. Evaluation of antioxidant potential of honey drops and honey lozenges. *Food Chem. Adv.* **2022**, *1*, 100013. [CrossRef]

90. Meda, A.; Lamien, C.E.; Romito, M.; Millogo, J.; Nacoulma, O.G. Determination of the total phenolic, flavonoid and proline contents in Burkina Fasan honey, as well as their radical scavenging activity. *Food Chem.* **2005**, *91*, 571–577. [CrossRef]

91. Pešić, M.B.; Pešić, M.M.; Bezbradica, J.; Stanojević, A.B.; Ivković, P.; Milinčić, D.D.; Demin, M.; Kostić, A.Ž.; Dojčinović, B.; Stanojević, S.P. Okara-enriched gluten-free bread: Nutritional, antioxidant and sensory properties. *Molecules* **2023**, *28*, 4098. [CrossRef]

92. Wilczyńska, A. Phenolic Content and Antioxidant Activity of Different Types of Polish Honey—A Short Report. *Pol. J. Food Nutr. Sci.* **2010**, *60*, 309–313. Available online: http://journal.pan.olsztyn.pl (accessed on 1 July 2024).

93. Gülçin, I. Antioxidant activity of food constituents: An overview. *Arch. Toxicol.* **2012**, *86*, 345–391. [CrossRef]

94. Predescu, C.; Papuc, C.; Nicorescu, V. Antioxidant activity of sunflower and meadow honey. *Sci. Work. Ser. C Vet. Med.* **2015**, *LXI*, 45–50. Available online: https://www.cabidigitallibrary.org/doi/pdf/10.5555/20163206566 (accessed on 16 July 2024).

95. Ulusoy, E.; Kolayli, S.; Sarikaya, A.O. Antioxidant and antimicrobial activity of different floral origin honeys from Turkiye. *J. Food Biochem.* **2010**, *34*, 321–335. [CrossRef]

96. Karaçelik, A.A.; Sahin, H. Determination of enzyme inhibition and antioxidant activity in some chestnut honeys. *Foods Raw Mater.* **2018**, *6*, 210–218. [CrossRef]

97. Felsner, M.L.; Cano, C.B.; Matos, J.R.; Almeida-Muradian, L.B.; Bruns, R.E. Optimization of thermogravimetric analysis of ash content in honey. *J. Braz. Chem. Soc.* **2004**, *15*, 797–802. [CrossRef]

98. Doner, L.W. Honey. In *Encyclopedia of Food Sciences and Nutrition*, 2nd ed.; Caballero, B., Finglas, P.M., Trugo, L.C., Eds.; Academic Press: London, UK, 2020; pp. 3125–3130. [CrossRef]

99. Popa, M.; Bostan, R.; Popa, D. Honey-marker of environmental pollution. Case study-the Transylvania Region, Romania. *J. Environ. Prot. Ecol.* **2013**, *14*, 273–280.

100. Maté, A.P. Characterization of Artisanal Honeys from Castilla Y León (Spain). Ph.D. Thesis, Department of Biotechnology and Food Science, Faculty of Sciences, University of Burgos, Burgos, Spain, 2016; pp. 23–78.

101. Oroian, M.; Ropciuc, S. Botanical authentication of honeys based on Raman spectra. *Food Meas.* **2017**, *12*, 545–554. [CrossRef]

102. Anjos, O.; Santos, A.J.A.; Paixão, V.; Estevinho, L.M. Physicochemical characterization of Lavandula spp. honey with FT-Raman spectroscopy. *Talanta* **2018**, *178*, 43–48. [CrossRef]

103. de Oliveira, L.F.C.; Colombara, R.; Edwards, H.G.M. Fourier transform Raman spectroscopy of honey. *Appl. Spectrosc.* **2002**, *56*, 306–311. [CrossRef]

104. Tahir, H.E.; Xiaobo, Z.; Zhihua, L.; Jiyong, S.; Zhai, X.; Wang, S.; Mariod, A.A. Rapid prediction of phenolic compounds and antioxidant activity of Sudanese honey using Raman and Fourier transform infrared (FT-IR) spectroscopy. *Food Chem.* **2017**, *226*, 202–211. [CrossRef]

105. Corvucci, F.; Nobili, L.; Melucci, D.; Grillenzoni, F.V. The discrimination of honey origin using melissopalynology and Raman spectroscopy techniques coupled with multivariate analysis. *Food Chem.* **2015**, *169*, 297–304. [CrossRef]

106. Kizil, R.; Irudayaraj, J.; Seetharaman, K. Characterization of irradiated starches by using FT-Raman and FTIR spectroscopy. *J. Agric. Food Chem.* **2002**, *50*, 3912–3918. [CrossRef]

107. Pierna JA, F.; Abbas, O.; Dardenne, P.; Baeten, V. Discrimination of Corsican honey by FT-Raman spectroscopy and chemometrics. *Biotechnol. Agron. Soc. Environ.* **2011**, *15*, 75–84.

108. Madgas, D.A.; Guyon, F.; Berghian-Grosan, C.; Molnar, M.C. Challenges and a step forward in honey classification based on Raman spectroscopy. *Food Control* **2021**, *123*, 107769. [CrossRef]
109. Ferreira, E.L.; Lencioni, C.; Benassi, M.T.; Barth, M.O.; Bastos, D.H.M. Descriptive sensory analysis and acceptance of stingless bee honey. *Food Sci. Technol. Int.* **2009**, *15*, 251–258. [CrossRef]

*foods*

MDPI

Article

Chemical and Functional Characteristics of Strawberry Tree (*Arbutus unedo* L.) Honey from Western Greece

Chrysoula Tananaki, Dimitrios Kanelis *, Vasilios Liolios, Maria Anna Rodopoulou and Fotini Papadopoulou

Laboratory of Apiculture-Sericulture, Aristotle University of Thessaloniki, Aristotle Farm, 57001 Thermi, Greece; tananaki@agro.auth.gr (C.T.); vliolios@agro.auth.gr (V.L.); rodopoum@agro.auth.gr (M.A.R.); foteinpi@agro.auth.gr (F.P.)
* Correspondence: dkanelis@agro.auth.gr; Tel.: +30-2310-991752

Abstract: Strawberry tree honey (*Arbutus unedo* L.) is a rare monofloral honey type with unique characteristics, mainly produced in the Mediterranean region. Despite its distinct qualities, limited research on its physicochemical and biological properties, coupled with the absence of specific legislative standards, hinder its market potential. For this reason, in the present study, we analyzed strawberry tree honey samples collected from beekeepers in Western Greece, focusing on physicochemical properties (moisture, electrical conductivity, HMF, diastase activity, color, pH, acidity), total phenolic content, antioxidant activity, carbohydrate composition, and phenolic compounds profile. The results revealed high moisture content ($19.2 \pm 1.9\%$) and electrical conductivity (0.784 ± 0.132 mS cm^{-1}), low diastase activity (9.6 ± 3.8 DN), and a strong crystallization tendency (1.01). Additionally, the honey exhibited elevated levels of total phenolic content (1169.9 ± 323.8 mg GAE kg^{-1} honey) and total antioxidant activity (10.98 ± 2.42 mmol Fe^{2+} kg^{-1} honey), compared to other blossom honeys, with homogentisic acid emerging as the dominant phenolic compound. These findings highlight the potential of strawberry tree honey as a high-value product, contributing to its enhanced market promotion.

Keywords: strawberry tree honey; physicochemical characteristics; total phenolic content; total antioxidant activity; carbohydrate profile; phenolic compounds profile

Academic Editors: Liming Wu and Qiangqiang Li

Received: 4 April 2025
Revised: 16 April 2025
Accepted: 22 April 2025
Published: 23 April 2025

Citation: Tananaki, C.; Kanelis, D.; Liolios, V.; Rodopoulou, M.A.; Papadopoulou, F. Chemical and Functional Characteristics of Strawberry Tree (*Arbutus unedo* L.) Honey from Western Greece. *Foods* **2025**, *14*, 1473. https://doi.org/10.3390/foods14091473

1. Introduction

Nowadays, honey is considered a basic part in consumers' diets as they embrace healthier lifestyles. The biological properties of honey, such as antioxidant and antimicrobial activities, as well as its sensory characteristics (e.g., color, taste, texture) have been found to depend mainly on its botanical origin, leading to an increase in the demand of monofloral honeys. Combined with the limited geographical areas and short flowering periods required for their production, monofloral honeys typically command higher prices compared to blends. For these reasons, beekeepers around the world tend to produce monofloral honeys (e.g., thyme honey, pine honey, etc.), many of which stand out for their unique characteristics. Furthermore, to compete with low-cost imported honey, beekeepers are increasingly focusing on producing high-quality monofloral honey types to meet market demands.

Strawberry tree (*Arbutus unedo*) honey, a type of honey produced along the Mediterranean, known for its characteristic bitter taste and persistent aroma, has seen increasing demand and consumption in recent years [1,2]. Regardless of its great value, there is limited research on its nutritional and biological benefits, as well as its unique organoleptic properties, which prevents it from receiving the recognition it truly deserves. This type of honey has been mainly explored in Sardinia [2–6], while few papers exist in the literature concerning its pollen profile [1,7], physicochemical characterization [6,7], and biological properties [6,8–11].

Several researchers have focused on the distinctive phenolic and volatile compounds present in strawberry tree honey. Regarding phenolic compounds, homogentisic acid has been widely reported as a floral origin marker [1,9,12], while Tuberoso et al. [1] additionally identified (±)-2-cis,4-trans-abscisic acid (c,t-ABA), (±)-2-trans,4-trans-abscisic acid (t,t-ABA), and unedone (2-(1,2-dihydroxypropyl)-4,4,8-trimethyl-1-oxaspiro [2.5] oct-7-en-6-one) as significant components, with unedone also confirmed by Montoro et al. [13]. Concerning volatile compounds, Bianchi et al. [3] highlighted α-isophorone, β-isophorone, and 4-oxoisophorone as markers of strawberry tree honey, while Osés et al. [10] proposed 2,6,6-trimethyl-4-oxo-2-cyclohexen-1-carboxaldehyde, 3,4,5-trimethylphenol, and 2-hydroxycyclopent-2-en-1-one as potential markers.

Greece's diverse climate supports the production of various monofloral honey types, including strawberry tree honey, especially in the western regions; however, it is not produced by many beekeepers. Furthermore, consumer unawareness and the absence of legislative criteria hinder the market promotion of this type of honey, leaving significant quantities unsold. Considering all the above, in the present study, monofloral strawberry tree honey samples were collected from beekeepers in Western Greece and analyzed for their physicochemical and biological properties with the aim to highlight the unique quality characteristics of strawberry tree honey and to further emphasize its nutritional significance. The findings could potentially contribute to the development of legislative standards, helping to define and elevate the quality of strawberry tree honey.

2. Materials and Methods

2.1. Collection of Strawberry Tree Honey Samples

Strawberry tree honey samples were collected from collaborated beekeepers located in Western Greece (Regional Units of Ioannina, Arta, Thesprotia). Beekeepers were provided with instructions to apply the appropriate beekeeping practices in order to collect representative honey samples. The samples, after their collection, were sent to the Laboratory of Apiculture, Sericulture of Aristotle University of Thessaloniki (AUTH) and kept in a freezer (−18 °C) until their analysis. To certify their botanical origin, the samples were at first studied for their pollen and sensory characteristics. The qualitative pollen analysis was performed using the method established by the International Commission of Bee Botany described by Von der Ohe et al. [14]. Furthermore, a panel of 6 experts tested the honey samples and confirmed their botanical origin, by approving or disapproving the monofloral nature of each sample with a yes or no response, based on its color, taste, odor, and aroma [15]. In total, 37 monofloral strawberry tree honey samples were analyzed.

2.2. Analyses of Physicochemical Characteristics

The physicochemical analyses of the collected strawberry tree honey samples were conducted following the methods outlined by Bogdanov et al. [16].

The water content was analyzed using an ATAGO refractometer (ATAGO Co., Ltd., Tokyo, Japan, HHR-2N), while for the determination of the electrical conductivity, an amount calculated using the formula m = 500/(100 − water content (%)) was diluted with deionized water to reach a total volume of 25 mL. The WTW conductometer (Cond315i, WTW GmbH, Weilheim, Germany) was used, the electrode of which was immersed in the solution, providing the value adjusted to 20 °C. The hydroxymethylfurfural (HMF) content was identified by measuring its UV absorbance at 284 nm, and subtracting the background absorbance at 336 nm. Diastase activity was measured through the spectrophotometric method, using a buffered solution of soluble starch and honey incubated in a thermostatic bath at 40 °C.

The pH of the honey solutions was identified using a Nahita 902/4 pH meter (Auxilab S.L., Beriáin, Spain), while for the determination of free acidity, the solutions were titrated with a solution of 0.05 mol L^{-1} sodium hydroxide (NaOH, Lach-Ner, Neratovice, Czech Republic), until reaching the pH value of 8.3.

The color was determined using a honey color analyzer (Hanna, HI-83221, Nușfalău, Romania) in Pfund scale and a Konica Minolta colorimeter (Konica Minolta, CR-410, Tokyo, Japan). To measure the CIE L* a* b* color coordinates, the colorimeter was calibrated using a white standard plate with color coordinates of Υ = 85.8, x = 0.3192, and y = 0.3369. The L* coordinate indicates brightness, ranging from L* = 0 (black) to L* = 100 (colorless). The a* coordinate represents the green/red spectrum, with a* > 0 indicating red and a* < 0 indicating green. The b* coordinate reflects the blue/yellow spectrum, where b* > 0 signifies yellow and b* < 0 signifies blue.

2.3. Evaluation of Biological Properties In Vitro

The samples were also analyzed for their antioxidant potential by measuring their total phenolic content (TPC) (Meda et al., 2005) [17] and total antioxidant activity through the Ferric Reducing Power (FRAP) Assay (Benzie and Strain, 1999) [18].

For the determination of TPC, the Folin–Ciocalteu colorimetric method was applied, where honey samples diluted in distilled water (1:10) were filtered and 500 µL of the resulting solution was mixed with 2.5 mL of 0.2 N Folin–Ciocalteu phenol reagent (2 N) (Merck, Darmstadt, Germany) and kept for 5 min at room temperature. Then, 2 mL of 75 g L^{-1} sodium carbonate (Na_2CO_3) ($\geq$99.5%, Merck, Darmstadt, Germany) was added and the mixture was incubated for 2 h in the dark at room temperature. The absorbance was measured at 760 nm (Genesys 10S UV-Vis, Thermo Fisher Scientific, Waltham, MA, USA) using a methanol blank as a reference ($\geq$99.8%, Chem-Lab, Zedelgem, Belgium). The analysis was performed in duplicate. For the calibration curve, standard solutions of gallic acid (10–400 mg L^{-1}) were used, while the results were expressed as mg gallic acid equivalents (GAEs) per 100 g^{-1} honey.

For the assessment of total antioxidant activity, the honey samples were diluted with distilled water in a (1:5) (w/v) ratio, which was stirred for 20 min. Then, 3 mL from the FRAP solution was mixed with 100 µL of the honey solution. The fresh FRAP reagent contained a solution of 300 mM $CH_3COONa - CH_3COOH$ ($\geq$99%, Chem-Lab, Zedelgem, Belgium) buffer (pH = 3.6), a solution of 10 mM TPTZ (2,4,6-tripyridyl-s-triazine) ($\geq$98.0%, Merck, Darmstadt, Germany) in 40 mM HCl (3 M, Chem-Lab, Zedelgem, Belgium), and an aquatic solution of 20 mM $FeCl_3 \times 6H_2O$ (>99%, Chem-Lab, Zedelgem, Belgium), in a ratio of 10:1:1. The samples were stirred and transferred to a water bath at 37 °C for 4 min. Absorbance was measured at 593 nm, using water (100 µL) with 3 mL of FRAP as the reference solution. The analysis was performed in duplicate. Standard solutions of iron (II) sulfate heptahydrate (concentration: 50–2000 µM) (>99.7%, Chem-Lab, Zedelgem, Belgium) were used for the calibration curves, while the results were expressed as millimoles of iron (II) sulfate heptahydrate per kg of honey (mmol Fe^{2+} kg^{-1} honey).

2.4. Carbohydrate and Phenolic Compounds Profile Analysis

The method of Bogdanov et al. [19] was used in order to analyze the carbohydrate profile using high-performance liquid chromatography (HPLC) equipped with a refractive index detector system (RID) (Agilent, 1200 series, Santa-Clara, CA, USA). An amount of 5 g of honey samples was dissolved in a methanol:water (25:75, v/v) (HPLC grade, $\geq$99.9%, Chem-Lab, Zedelgem, Belgium) solution to a final volume of 50 mL and filtered through a

0.45 μm disposable syringe filter prior to injection. Carbohydrate separation was conducted using two Zorbax Carbohydrate Analysis columns (4.6 mm ID × 150 mm × 5 μm) (Agilent Technologies, Inc. Headquarters, Santa-Clara, CA, USA) arranged in series and protected by a guard column (NH_2 Guard Cartridge, 4.6 mm × 12.5 m) (Agilent Technologies, Inc. Headquarters, Santa-Clara, CA, USA). The mobile phase consisted of an acetonitrile:water (75:25, *v/v*) (HPLC grade, ≥99.9%, Chem-Lab, Zedelgem, Belgium) solution at a flow rate of 1.8 mL min^{-1}. The column and detector were maintained at a constant temperature of 35 °C, with an injection volume of 10 μL. Quantification was performed using five-point calibration curves for each carbohydrate compound. Sixteen carbohydrate compounds were studied: D(−)-fructose, D-(+)-glucose, D(+)-sucrose, D-maltose monohydrate, D-(+)-turanose, D-(+)-trehalose dehydrate, isomaltose, D(+)-maltotriose, D-(+)-melezitose hydrate, erlose, D-raffinose pentahydrate, melibiose, D-panose, maltulose, maltotetraose and isomaltotriose, all of HPLC grade (Merck, Darmstadt, Germany).

To determine the phenolic compounds profile, the methods of Can et al. [20] and Akuyz et al. [21] were followed, with some modifications. Phenolic acids homogentisic, protocatechuic, caffeic, syringic, p-coumaric, ellagic, as well as flavonoids catechin, epi-catechin, rutin, quercitrin, quercetin, chrysin (≥98%, Merck, Darmstadt, Germany) were analyzed using a high liquid chromatography with a diode-array detector (HPLC-DAD) (Agilent Technologies, Inc. Headquarters, Santa-Clara, CA, USA). The honey samples (15 g) were mixed with 50 mL of HPLC-grade methanol with continuous stirring until fully dissolved. The solution was filtered and transferred to 50 mL volumetric flasks, where the volume was adjusted with methanol. The prepared solutions were transferred into beakers, sealed with parafilm, and stored at −80 °C for 24 h. After freezing, the samples were lyophilized and subjected to liquid−liquid extraction. Specifically, 20 mL of ultrapure water (Merck, Darmstadt, Germany) and 20 mL of diethyl ether (≥99%, Chem-Lab, Zedel-gem, Belgium) were added to each sample. The mixtures were left at room temperature until two distinct layers formed: a diethyl ether phase (up) and a honey−water phase (down). The ether phase was carefully collected into a beaker, stored again at −80 °C for 24 h, and lyophilized a second time. The remaining extract was dissolved in methanol and transferred to HPLC vials, to which 70 μL of internal standard solution of propylparaben (100 ppm) (Extrasynthese, Lyon, France) was added. The final volume in each vial was adjusted to 1 mL.

The samples were analyzed in HPLC-DAD using a C18 column (4.6 × 150 mm, 3 μm) on an Agilent 1200 liquid chromatography system (Agilent Technologies, Inc. Headquar-ters, Santa-Clara, CA, USA) with a gradient elution program. Two solvent systems were employed: Solvent A (80% acetonitrile in methanol) and Solvent B (2% acetic acid in ultra-pure water). The elution progressed from high polarity and low pH to low polarity and high pH. The gradient program was the following: 0–2 min, 95% B; 2–8 min, 95–90% B; 8–11 min, 90–85% B; 11–13 min, 85–75% B; 13–17 min, 75–70% B; 17–30 min, 70–65% B; 30–33 min, 65–0% B; 33–38 min, 0–0% B; 38–40 min, 95% B; and 40–48 min, 95% B. The injection volume was set to 50 μL, with a column temperature maintained at 30 °C. The flow rate was 1 mL min^{-1}, and detection was carried out at a wavelength of 290 nm. All solvents were of HPLC grade. Quantification was performed using five-point calibration curves for each phenolic compound.

2.5. Statistical Analysis

The statistical analysis was performed in SPSS v.24.0 software (Chicago, IL, USA) and expressed as mean values ± standard deviations. The level of significance was set at a = 0.05.

3. Results and Discussion

3.1. Physicochemical Characteristics

The botanical origin of the collected samples was at first confirmed by the combination of melissopalynological and sensory analyses. Strawberry tree honey is normally under-represented, while there are no legislative criteria established regarding its pollen percentage. Also, the coexistence of overrepresented pollens may cause a wide variation in the percentages of *Arbutus* pollen [1]. The range of *Arbutus* pollen grains was 4–59%. It should be noted that pollen grains of *Hedera helix* were also found in the samples, with percentages ranging from 31 to 84%. Pollen grains from *Asparagus* sp., *Erica manipuliflora*, Fabaceae, and *Trifolium* sp. were also detected, while regarding nectarless plants, pollen grains from *Quercus* sp. and Cyclamen type were mostly found. Persano Oddo et al. [7] pointed out higher limits for Arbutus pollen grains (>8%), which could be attributed to the different geographical origin of the specific honey type.

After the confirmation of their botanical origin, the samples were analyzed for their physicochemical characteristics. The results are given in Table 1.

Table 1. Physicochemical characteristics (mean values, standard deviations, min, max) of strawberry tree honey.

Physicochemical Parameters	Moisture (%)	Electrical Conductivity (mS cm^{-1})	HMF (mg kg^{-1})	Diastase Activity (DN)	pH	Acidity (meq kg^{-1})
Mean value ± Standard deviation	19.2 ± 1.9	0.784 ± 0.132	3.1 ± 3.3	9.6 ± 3.8	4.3 ± 0.16	34.1 ± 7.6
Min-Max	16.4–24.0	0.454–1.120	0.0–12.0	3.2–18.0	3.72–4.41	17.5–49.0

The collected samples exhibited a relatively high mean moisture content (19.2 ± 1.9%), with 38% exceeding the moisture limit (lower than 20%) established by the European legislation [22]. Electrical conductivity was also found in high levels (mean value: 0.784 ± 0.132 mS cm^{-1}), with 49% of the samples surpassing the threshold of 0.8 mS cm^{-1} which distinguishes blossom from honeydew honeys; however, this particular honey type is classified as an exception under the Honey Directive [22], regarding its conductivity. Additionally, acidity was notably high (average: 34.1 ± 7.6 meq kg^{-1}), while 30% of the samples exceeded 40 meq kg^{-1}, leading to restrains of their shelf life. Acidity is a useful criterion for the evaluation of fermentation and characterization of monofloral honeys, and according to the Honey Directive [22], it should be lower than 50 meq kg^{-1}. Regarding pH, all the samples were acidic, with a mean pH value of 4.03 ± 0.16. Moreover, the diastase activity was found to be relatively low (mean value: 9.6 ± 3.8 DN), and 38% of the samples had diastase values lower than 8 DN, suggesting strawberry tree honey to be added to the legislative exceptions [23], considering that the minimum legislative value of diastase activity is 8 DN, with exceptions including honeys with low enzyme content that require diastase activity above 3 DN and HMF content below 15 mg kg^{-1} (e.g., citrus honeys) [22]. The HMF content was below 15 mg kg^{-1} (3.1 ± 3.3 mg kg^{-1}), proving that the samples were fresh and unprocessed. Finally, the honey was classified as light brown on the Pfund scale, with an average value of 81.4 ± 22.8 mm Pfund, while most samples had color values between 60 and 80 mm Pfund (Figure 1a).

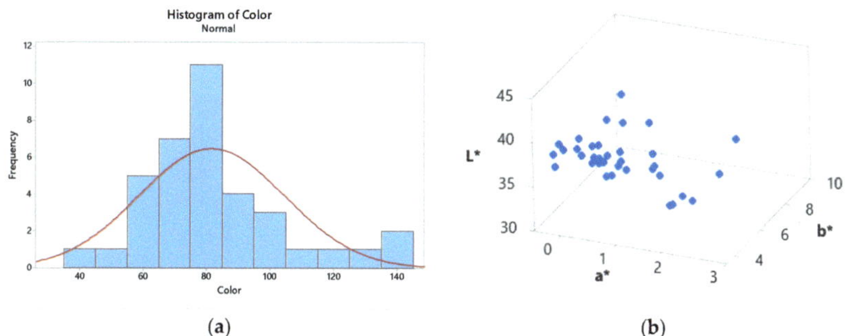

(a)   (b)

Figure 1. Color (mm Pfund) (**a**) and L*, a*, b* parameters (**b**) in strawberry tree honey samples (*n* = 37). L* represents the clarity (L* = 0 black and L* = 100 colorless), a* the green/red color component (a* > 0 red, a* < 0 green), and b* the blue/yellow color component (b* > 0 yellow, b* < 0 blue).

This classification aligns with the L* mean value of 36.80 ± 1.27, indicating predominantly yellow hues followed by green tones, as reflected in the b* and a* mean values (5.22 ± 1.09 and 0.66 ± 0.77, respectively), and as presented in Figure 1b. The yellow hues are mainly attributed to the presence of carotenoids and flavonoids in the nectar, while the green hues may be due to a high chlorophyll content [24].

The findings of the present study coincided with those reported by Persano Oddo et al. [7] regarding moisture content (18.9 ± 1.9%), electrical conductivity (0.740 ± 0.07 mS cm^{-1}), HMF levels (4.4 ± 3.2 mg kg^{-1}), and pH (4.2 ± 0.1) and by Rodopoulou et al. [25], as well (moisture content: 19.0 ± 1.7%, electrical conductivity: 0.740 ± 0.10 mS cm^{-1}, pH: 4.2 ± 0.1). However, Persano Oddo et al. [7] observed lower color values (70 ± 10 mm Pfund), lower diastase activity (5.2 ± 2.8 DN), and slightly higher acidity (39.6 ± 8.3 meq kg^{-1}). Castiglioni et al. [26] also found lower color values (70 ± 10 mm Pfund), compared to the results of this study. On the other hand, Petri and Tarola [6], in their analysis of five honey samples from Sicily and five from Sardinia, recorded higher HMF levels (11.1 ± 1.05 mg kg^{-1} in honeys from Sicily and 19.63 ± 4.65 mg kg^{-1} in honeys from Sardinia) but similar pH values (4.26 ± 0.10 and 4.24 ± 0.10) to those found in this study, while the acidity levels observed in the present research were more consistent with the honey samples from Sardinia (30.81 ± 3.63 meq kg^{-1}) than with those from Sicily (39.72 ± 3.11 meq kg^{-1}).

3.2. Antioxidant Properties

The collected honey samples were also analyzed for their total phenolic content (TPC) and their total antioxidant activity by applying the FRAP assay (Figure 2).

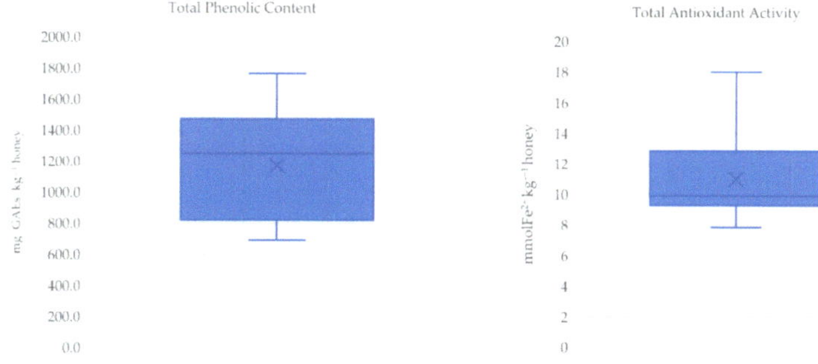

Figure 2. Total phenolic content and total antioxidant activity of strawberry tree honey.

The TPC values were relatively high, averaging 1169.9 $\pm$ 323.8 mg GAEs kg^{-1} honey. Petri and Tarola [6] found similar values (1035.3 mg GAEs kg^{-1} honey for honey from Sicily, 956.4 mg GAEs kg^{-1} honey for honey from Sardinia) along with Rosa et al. [8] (972 mg GAEs kg^{-1} honey), while Osés et al. [10] reported slightly higher values (1500 $\pm$ 300 mg GAEs kg^{-1} honey). Afrin et al. [4] observed lower total phenolic values (from 690 $\pm$ 100 mg GAEs kg^{-1} honey to 1000 $\pm$ 200 mg GAE kg^{-1} honey), and Castiglioni et al. [26] as well (850 $\pm$ 133 mg GAEs kg^{-1} honey). Furthermore, the mean value of total antioxidant activity was 10.98 $\pm$ 2.42 mmol Fe^{2+} kg^{-1} honey, similar to the values found by Rosa et al. [8] (11.7 $\pm$ 1.7 mmol Fe^{2+} kg^{-1} honey) and Tuberoso et al. [27] (12.0 $\pm$ 2.2 mmol Fe^{2+} kg^{-1} honey), while it was higher than that identified by Afrin et al. [4] (5.1–9.2 mmol Fe^{2+} kg^{-1} honey). In comparative studies, strawberry tree honey was found among the types with the highest antioxidant activity, attributing this potential to its amount of polyphenols [9,26].

3.3. Carbohydrate and Phenolic Compounds Profile

Carbohydrate profiling revealed an average fructose content of 36.99 $\pm$ 3.29% and glucose at 36.53 $\pm$ 2.34% (Figure 3), the sum of which (73.52%) is above 60%, the lower limit for blossom honeys [22].

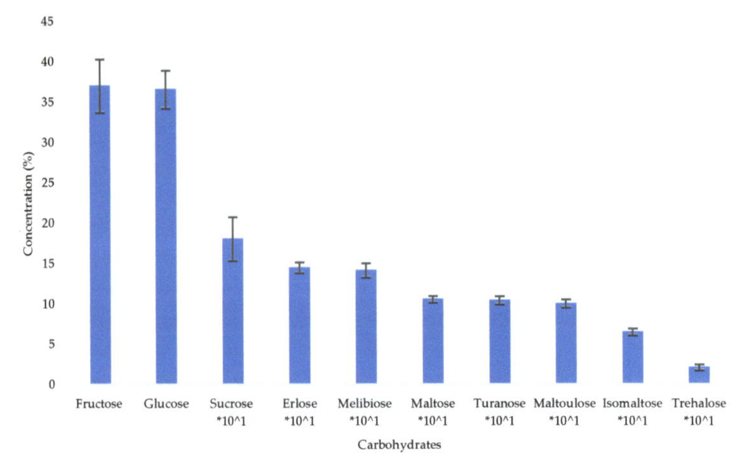

Figure 3. Carbohydrates mean concentration (%) in strawberry tree honey.

The sum of the main sugars fructose and glucose, as well as their ratio, have been found to affect the degree of crystallization in honey; honeys with a low fructose/glucose ratio (<1.11) tend to crystallize faster compared to honeys with a high fructose/glucose ratio [28]. In the present study, strawberry tree honey had a low fructose/glucose ratio (1.01), showing a fast rate of crystallization; therefore, care should be taken during its shelf storage. Additionally, sucrose levels were generally low; however, 22% of the samples exceeded the legislative limit of 5% [22]. This could be due to the late autumn collection of strawberry tree honey, which limits the time available for honey bees to process the nectar.

Erlose, melibiose maltose, turanose, and maltulose followed in lower percentages (average: 1.45 $\pm$ 0.64%, 1.41 $\pm$ 0.88%, 1.06 $\pm$ 0.39%, 1.04 $\pm$ 0.50%, and 1.00 $\pm$ 0.50%, respectively), while even lower mean concentrations were observed for isomaltose (0.65 $\pm$ 0.42%) and trehalose (0.21 $\pm$ 0.34%). From the carbohydrates analyzed, raffinose, maltotriose, panose, isomaltotriose, and maltotetraose were not detected. The results are in agreement with those presented by Persano Oddo et al. [7],who also found fructose ($37.2 \pm 2.4\%$), glucose ($32.1 \pm 1.1\%$), sucrose ($1.5 \pm 0.9\%$), maltose ($1.2 \pm 0.5\%$), isomaltose ($0.8 \pm 0.3\%$), and erlose.

Regarding phenolic compounds, the homogentisic acid was detected in higher concentrations (2681.1 ± 1645.5 µg 100 g^{-1} honey) (Figure 4), in agreement with other studies, proposing the specific compound as a potential origin marker [8,12].

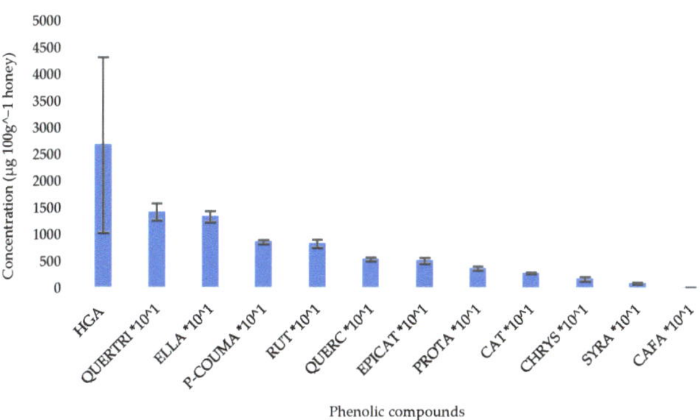

Figure 4. Phenolic compounds concentration in Arbutus honey samples from Western Greece (HGA: homogentisic acid; QUERTRI: quercitrin; ELLA: ellagic acid; p-COUMA: p-coumaric acid; RUT: rutin; QUERC: quercetin; EPICAT: epicatechin; PROTA: protocatechuic acid; CAT: catechin; CHRYS: chrysin; SYRA: syringic acid; CAFA: caffeic acid).

Rosa et al. [8] even reported that the homogentisic acid represented 50–60% of the total phenolic compounds in the honey, while they found that the phenolic compound itself showed a high antioxidant activity and a protective effect against the oxidation of LDL cholesterol, which makes an important contribution to the high antioxidant and antiradical properties of strawberry tree honey. Homogentisic acid has been also found to offer protection against light and oxidation stress, ROS and DPPH radical scavenging activities, and protecting human peripheral blood lymphocytes against irinotecan-induced cytogenetic damage [29]. The compounds ellagic acid, quercitrin, p-coumaric acid, and rutin followed in mean concentrations 134.15 ± 155.56 µg 100 g^{-1} honey, 134.10 ± 103.28 µg 100 g^{-1} honey, 86.82 ± 38.46 µg 100 g^{-1} honey, and 83.53 ± 73.68 µg 100 g^{-1} honey, respectively. The compounds caffeic acid, p-coumaric acid, syringic acid, ellagic acid, rutin, and chrysin are also noted in the study by Petri and Tarrola [6], while caffeic acid, p-coumaric acid, quercetin, and chrysin were detected in the study of Jurič et al. [30]. The comparison of the studied phenolic compounds in monofloral honey types with the results of other studies is difficult, not only due to the origin of the honey, but also due to the process applied during the extraction of phenolic compounds and their analysis. The results between studies can be comparable only when the same analysis method has been followed [31]. As for the phenolic compounds ellagic acid, quercitrin, p-coumaric acid, and rutin, their biological potential has been researched, confirming their role as antioxidant, antimicrobial, anti-allergic, anti-inflammatory, and anti-apoptotic agents [32–35].

The study focuses on strawberry tree honey, a unique and rare monofloral honey type, which has not been extensively researched compared to the more common honey types. The researches that exist so far mostly refer to the region of Italy-Sardinia. To our knowledge, this is the first time that a comprehensive study, including a large number of samples ($n = 37$) and the examination of different parameters (water content, electrical conductivity, HMF, diastase activity, color, pH, acidity, total phenolic content, total antioxidant activity, carbohydrate profile, phenolic compounds profile) was being conducted on Arbutus honey produced in Western Greece, providing data that could help in promoting regional honey production and supporting local beekeepers. The large number of samples aids in documenting deviations in certain physicochemical characteristics, which support the improvement in legislation regarding this type of honey. Moreover, most of the researches identified the main carbohydrates (fructose and glucose), while in the present study, besides the main carbohydrates, disaccharides and trisaccharides were also determined: D(+)-sucrose, D-maltose monohydrate, D-(+)-turanose, D-(+)-trehalose dehydrate, isomaltose, D(+)-maltotriose, D-(+)-melezitose hydrate, erlose, D-raffinose pentahydrate, melibiose, D-panose, maltulose, maltotetraose, and isomaltotriose, focusing on the identification of a carbohydrate profile, instead of specific carbohydrates. Finally, few studies exist that have detected specific phenolic compounds, besides the homogentisic acid, the results of which could be used in future clinical tests.

4. Conclusions

As consumer preferences increasingly shift toward distinctive food products, the results of the present study could support the promotion of strawberry tree honey, a monofloral honey type with distinct organoleptic characteristics. It was found that the specific honey type has light brown color, and is characterized by high moisture content and electrical conductivity, low diastase activity, and strong tendency to crystallize. It is also rich in phenolic compounds, with homogentisic acid being the most abundant. Significant amounts of the compounds quercitrin, ellagic acid, p-coumaric acid, and rutin were also detected, all of which are known for their beneficial effects on human health.

Author Contributions: Conceptualization, C.T., D.K. and V.L.; methodology, C.T., D.K. and V.L.; investigation, D.K., V.L., M.A.R. and F.P.; writing—original draft preparation, D.K., V.L., M.A.R. and F.P.; writing—review and editing, C.T., D.K. and V.L.; visualization, D.K., V.L., M.A.R. and F.P.; supervision, C.T. All authors have read and agreed to the published version of the manuscript.

Funding: The study is part of the project titled "Strawberry tree honey: Creation of a value chain, through highlighting its special characteristics, development of an innovative model for the promotion and optimization of the production process". The research project is co-funded by: (a) the European Agricultural Fund for Rural Development (EAFRD), (b) the Agricultural Development Program (RAP) 2014–2022, (c) the Partnership Agreement for the Development Framework (PA) 2014–2020, and (d) the Hellenic Ministry of Rural Development and Food. Project Number: M16ΣΥN2-00428.

Institutional Review Board Statement: Not applicable.

Informed Consent Statement: Not applicable.

Data Availability Statement: The original contributions presented in the study are included in the article, further inquiries can be directed to the corresponding author. Data supporting the reported results are stored at the Laboratory of Apiculture-Sericulture, AUTH.

Acknowledgments: We would like to thank Konstantinos Anastasiadis for contributing to the collection of the honey samples and communicating with the beekeepers of Western Greece.

Conflicts of Interest: The authors declare no conflicts of interest.

References

1. Tuberoso, C.I.G.; Bifulco, E.; Caboni, P.; Cottiglia, F.; Cabras, P.; Floris, I. Floral Markers of strawberry tree (*Arbutus unedo* L.) honey. *J. Agric. Food Chem.* **2010**, *58*, 384–389. [CrossRef] [PubMed]

2.   Deiana, V.; Tuberoso, C.; Satta, A.; Pinna, C.; Camarda, I.; Spano, N.; Ciulu, M.; Floris, I. Relationship between markers of botanical origin in nectar and honey of the strawberry tree (*Arbutus unedo*) throughout flowering periods in different years and in different geographical areas. *J. Apic. Res.* **2016**, *54*, 342–349. [CrossRef]

3.   Bianchi, F.; Careri, M.; Musci, M. Volatile norisoprenoids as markers of botanical origin of Sardinian strawberry-tree (*Arbutus unedo* L.) honey: Characterisation of aroma compounds by dynamic headspace extraction and gas chromatography–mass spectrometry. *Food Chem.* **2005**, *89*, 527–532. [CrossRef]

4.   Afrin, S.; Forbes-Hernandez, T.Y.; Gasparrini, M.; Bompadre, S.; Quiles, J.L.; Sanna, G.; Spano, N.; Giampieri, F.; Battino, M. Strawberry-Tree honey induces growth inhibition of human colon cancer cells and increases ROS generation: A Comparison with Manuka Honey. *Int. J. Mol. Sci.* **2017**, *18*, 613. [CrossRef]

5.   Floris, I.; Pusceddu, M.; Satta, A. The Sardinian Bitter Honey: From Ancient Healing Use to Recent Findings. *Antioxidants* **2021**, *10*, 506. [CrossRef]

6.   Preti, R.; Tarola, A.M. Influence of geographic origin on the profile and level of phenolic compounds in Italian strawberry tree (*Arbutus unedo* L.) honey. *J. Food Nut. Res.* **2022**, *61*, 352–360.

7.   Persano Oddo, L.; Piazza, M.G.; Sabatini, A.G.; Accorti, M. Characterization of unifloral honeys. *Apidologie* **1995**, *26*, 453–465. [CrossRef]

8.   Rosa, A.; Tuberoso, C.I.G.; Atzeri, A.; Melis, M.P.; Bifulco, E.; Dessì, M.A. Antioxidant profile of strawberry tree honey and its marker homogentisic acid in several models of oxidative stress. *Food Chem.* **2011**, *129*, 1045–1053. [CrossRef]

9.   Lovaković, B.T.; Lazarus, M.; Brčić Karačonji, I.; Juricab, K.; Živković Semren, T.; Lušić, D.; Brajenović, N.; Pelaić, Z.; Pizent, A. Multi-elemental composition and antioxidant properties of strawberry tree (*Arbutus unedo* L.) honey from the coastal region of Croatia: Risk-benefit analysis. *J. Trace Elem. Med. Biol.* **2018**, *45*, 85–92. [CrossRef]

10.   Osés, S.M.; Nieto, S.; Rodrigo, S.; Pérez, S.; Rojo, S.; Sancho, M.T.; Fernández-Muiño, M.Á. Authentication of strawberry tree (*Arbutus unedo* L.) honeys from southern Europe based on compositional parameters and biological activities. *Food Biosci.* **2020**, *38*, 100768. [CrossRef]

11.   Jurič, A.; Brčić Karačonji, I.; Gašić, U.; Milojković Opsenica, D.; Prđun, S.; Bubalo, D.; Lušić, D.; Vahčić, N.; Kopjar, N. Protective Effects of *Arbutus unedo* L. Honey in the Alleviation of Irinotecan-Induced Cytogenetic Damage in Human Lymphocytes—An In Vitro Study. *Int. J. Mol. Sci.* **2023**, *24*, 1903. [CrossRef] [PubMed]

12.   Cabras, P.; Angioni, A.; Tuberoso, C.; Floris, I.; Reniero, F.; Guillou, C.; Ghelli, S. Homogentisic acid: A phenolic acid as a marker of strawberry-tree (*Arbutus unedo*) honey. *J. Agric. Food Chem.* **1999**, *47*, 4064–4067. [CrossRef]

13.   Montoro, P.; D'Urso, G.; Kowalczyk, A.; Tuberoso, C.I.G. LC-ESI/LTQ-Orbitrap-MS Based Metabolomics in evaluation of bitter taste of *Arbutus unedo* honey. *Molecules* **2021**, *26*, 2765. [CrossRef]

14.   Von Der Ohe, W.; Persano Oddo, L.; Piana, M.L.; Morlot, M.; Martin, P. Harmonized methods of Melissopalynology. *Apidologie* **2004**, *35*, 18–25. [CrossRef]

15.   Piana, M.L.; Persano Oddo, L.; Bentabol, A.; Bruneau, E.; Bogdanov, S.; Guyot-Declerck, C. Sensory analysis applied to honey: State of the art. *Apidologie* **2004**, *35*, 26–37. [CrossRef]

16.   Bogdanov, S.; Ruoff, K.; Oddo, L.P. Physico-chemical methods for the characterization of unifloral honey: A review. *Apidologie* **2004**, *35*, 275–282. [CrossRef]

17.   Meda, A.; Lamien, C.E.; Romito, M.; Millogo, J.; Nacoulma, O.G. Determination of the total phenolic, flavonoid and proline contents in Burkina Fasan honey, as well as their radical scavenging activity. *Food Chem.* **2005**, *91*, 571–577. [CrossRef]

18.   Benzie, I.F.F.; Strain, J.J. Ferric reducing/antioxidant power assay: Direct measure of total antioxidant activity of biological fluids and modified version for simultaneous measurement of total antioxidant power and ascorbic acid concentration. *Methods Enzymol.* **1999**, *299*, 15–27. [CrossRef]

19.   Bogdanov, S. Nature and Origin of the Antibacterial Substances in Honey. *LWT-Food Sci. Technol.* **1997**, *30*, 748–753. [CrossRef]

20.   Can, Z.; Yildiz, O.; Sahin, H.; Akyuz Turumtay, E.; Silici, S.; Kolayli, S. An investigation of Turkish honeys: Their physico-chemical properties, antioxidant capacities and phenolic profiles. *Food Chem.* **2015**, *180*, 133–141. [CrossRef]

21.   Akyuz, E.; Şahin, H.; Islamoglu, F.; Kolayli, S.; Sandra, P. Evaluation of Phenolic Compounds in *Tilia rubra* Subsp. *Caucasica* by HPLC-UV and HPLC-UV-MS/MS. *Int. J. Food Prop.* **2014**, *17*, 331–343. [CrossRef]

22.   European Economic Community EEC Council directive of 20 December 2001 relating to honey. *OJEC* **2002**, *110*, 47–50.

23.   Thrasyvoulou, A.; Tananaki, C.; Goras, G.; Karazafiris, E.; Dimou, M.; Liolios, V.; Kanelis, D.; Gounari, S. Legislation of honey criteria and standards. *J. Apic. Res.* **2018**, *57*, 88–96. [CrossRef]

24.   Flanjak, I.; Kenjerić, D.; Bubalo, D.; Primorac, L. Characterisation of selected Croatian honey types based on the combination of antioxidant capacity, quality parameters, and chemometrics. *Eur. Food Res. Technol.* **2016**, *242*, 467–475. [CrossRef]

25.   Rodopoulou, M.-A.; Tananaki, C.; Kanelis, D.; Liolios, V.; Dimou, M.; Thrasyvoulou, A. A chemometric approach for the differentiation of 15 monofloral honeys based on physicochemical parameters. *J. Sci. Food Agric.* **2021**, *102*, 139–146. [CrossRef]

26.   Castiglioni, S.; Stefano, M.; Astolfi, P.; Carloni, P. Chemometric approach to the analysis of antioxidant properties and colour of typical Italian monofloral honeys. *Int. J. Food Sci. Technol.* **2017**, *52*, 1138–1146. [CrossRef]

27. Tuberoso, C.I.G.; Boban, M.; Bifulco, E.; Budimir, D.; Pirisi, F.M. Antioxidant capacity and vasodilatory properties of Mediterranean food: The case of Cannonau wine, myrtle berries liqueur and strawberry-tree honey. *Food Chem.* **2013**, *140*, 686–691. [CrossRef] [PubMed]

28. Da Silva, P.M.; Gauche, C.; Gonzaga, L.V.; Costa, A.C.O.; Fett, R. Honey: Chemical composition, stability and authenticity. *Food Chem.* **2016**, *196*, 309–323. [CrossRef]

29. Jurič, A.; Brčić Karačonji, I.; Kopjar, N. Homogentisic acid, a main phenolic constituent of strawberry tree honey, protects human peripheral blood lymphocytes against irinotecan-induced cytogenetic damage in vitro. *Chem. Biol. Int.* **2021**, *349*, 109672. [CrossRef]

30. Jurič, A.; Brčić Karačonji, I.; Žunec, S.; Katić, A.; Gašić, U.; Milojković Opsenica, D.; Kopjar, N. Protective role of strawberry tree (*Arbutus unedo* L.) honey against cyto/genotoxic effects induced by ultraviolet B radiation in vitro. *J. Apic. Res.* **2022**, *63*, 513–522. [CrossRef]

31. Escriche, I.; Kadar, M.; Juan-Borrás, M.; Domenech, E. Suitability of antioxidant capacity, flavonoids and phenolic acids for floral authentication of honey. Impact of industrial thermal treatment. *Food Chem.* **2014**, *142*, 135–143. [CrossRef] [PubMed]

32. Chen, P.; Chen, F.; Zhou, B. Antioxidative, anti-inflammatory and anti-apoptotic effects of ellagic acid in liver and brain of rats treated by D-galactose. *Sci. Rep.* **2018**, *8*, 1465. [CrossRef] [PubMed]

33. Chen, J.; Li, G.; Sun, C.; Peng, F.; Yu, L.; Chen, Y.; Tan, Y.; Cao, X.; Tang, Y.; Xie, X.; et al. Chemistry, pharmacokinetics, pharmacological activities, and toxicity of Quercitrin. *Phytother. Res.* **2022**, *36*, 1545–1575. [CrossRef] [PubMed]

34. Pei, K.; Ou, J.; Huang, J.; Ou, S. p-Coumaric acid and its conjugates: Dietary sources, pharmacokinetic properties and biological activities. *J. Sci. Food Agric.* **2016**, *96*, 2952–2962. [CrossRef]

35. Negahdari, R.; Bohlouli, S.; Sharifi, S.; Maleki Dizaj, S.; Rahbar Saadat, Y.; Khezri, K.; Jafari, S.; Ahmadian, E.; Gorbani Jahandizi, N.; Raeesi, S. Therapeutic benefits of rutin and its nanoformulations. *Phytother. Res.* **2021**, *35*, 1719–1738. [CrossRef]

Article

The Use of Fluorescence Spectrometry Combined with Statistical Tools to Determine the Botanical Origin of Honeys

Aleksandra Wilczyńska * and Natalia Żak

Department of Quality Management, Gdynia Maritime University, ul. Morska 81-87, 81-225 Gdynia, Poland; n.zak@wznj.umg.edu.pl
* Correspondence: a.wilczynska@wznj.umg.edu.pl

Abstract: At a time when the botanical origin of honey is being increasingly falsified, there is a need to find a quick, cheap and simple method of identifying its origin. Therefore, the aim of our work was to show that fluorescence spectrometry, together with statistical analysis, can be such a method. In total, 108 representative samples with 10 different botanic origins (9 unifloral and 1 multifloral), obtained in 2020–2022 from local apiaries, were analyzed. The fluorescence spectra of those samples were determined using a F-7000 Hitachi fluorescence spectrophotometer, Tokyo, Japan. It is shown that each honey variety produces a unique emission spectrum, which allows for the determination of its botanical origin. Taking into account the difficulties in analyzing these spectra, it was found that the most information regarding botanical differences and their identification is provided by synchronous cross-sections of these spectra obtained at $\Delta\lambda = 100$ nm. In addition, this analysis was supported by discriminant and canonical analysis, which allowed for the creation of mathematical models, allowing for the correct classification of each type of honey (except dandelion) with an accuracy of over 80%. The application of the method is universal (in accordance with the methodology described in this paper), but its use requires the creation of fluorescence spectral matrices (EEG) characteristic of a given geographical and botanical origin.

Keywords: authenticity of honey; fluorescence spectra (EEM); method of identification; varieties and origin of honey

---

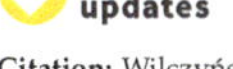

Citation: Wilczyńska, A.; Żak, N. The Use of Fluorescence Spectrometry Combined with Statistical Tools to Determine the Botanical Origin of Honeys. *Foods* **2024**, *13*, 3303. https://doi.org/10.3390/foods13203303

Academic Editor: Oscar Núñez Burcio

Received: 29 August 2024
Revised: 30 September 2024
Accepted: 15 October 2024
Published: 18 October 2024

1. Introduction

According to its definition, honey is a product produced by honeybees, *Apis mellifera*, from nectar or honeydew and occupies a special place in the human diet [1]. Its role in human nutrition is due to its rich taste and nutritional effects on the human body. The health-promoting properties of honey are closely related to its chemical composition, which depends mainly on the botanical origin of the honey (honey variety) [2,3]. Honey comes in many different varieties, depending on the source of the nectar honey (nectar and honeydew honey), the botanical origin and the geographical location [3]. The classification of nectar honey into a given variety depends on the plant pollen that is dominant. For example, in Central Europe, the most common varieties of honey are rapeseed, lime, dandelion, goldenrod, buckwheat and honeydew (from honeydew from deciduous and coniferous trees), as well as less common ones, such as phacelia, raspberry and clover. In other geographical zones, eucalyptus, chestnut, rose, lavender, orange and other honeys characteristic of the vegetation occurring in a given geographical zone are found. Honey in which no pollen predominates is classified as multifloral. Each of these honeys, depending on the variety, production method and origin, will have different properties.

The specific taste and health-promoting properties of honey contribute to its high price. The desire to increase profits encourages producers or traders to adulterate honey. Adulteration is carried out, for example, by mixing honeys of different varieties, mislabeling them, adding sugar syrups, repeatedly heating honeys, and other processes used intentionally by

---

producers [4]. However, recently, the most frequent fraud has been the mislabeling of the botanical origin of honeys [5]. In order to verify the quality of honeys and confirm their authenticity, a number of research methods are used. Mandatory methods for assessing the quality of honeys are usually described in legal acts. In Poland, this is the Regulation of the Minister of Agriculture and Rural Development, from 29 May 2015, amending the regulation on detailed requirements for the commercial quality of honey (Journal of Laws 2015, item 850). The only method recommended in legal acts to determine the botanical origin of honeys is pollen analysis. However, both in Poland and around the world, other, non-standard methods for assessing honey quality parameters are used, which can be used to determine not only the elementary chemical composition of individual honeys, but also the biological activity of honeys and their components [5–12]. As a result of the analysis of the literature on the assessment of the quality and authenticity of honeys, it was found that it is impossible to indicate repeatable, characteristic parameters, the value of which would allow for the unambiguous determination of the affiliation of a given sample of a specific honey variety. Various methods that have been used so far require not only financial outlays and time, but also confirmation of the accuracy of their result with alternative measurements. In many countries, compounds characteristic of a specific honey variety are sought, or the chemical profiles of a specific class of natural products are created (so-called "fingerprints" of individual honey varieties), but the results of these works are still unsatisfactory [4,5,12–15]. It can therefore be assumed that taking action to develop a reliable, fast and cheap method of assessing the quality and authenticity of honey will also allow for the detection and prevention of adulterated or lower-quality honey being introduced into circulation.

The use of excitation–emission matrices (EEMs) is one of two commonly used methods for measuring fluorescence spectra in the field of food safety and quality [16,17].

Although fluorescence spectroscopy techniques are widely used in the field of food quality and safety, they still have some limitations [18]. For example, Ruof et al. have been conducting research on the use of fluorescence-based methods for assessing the quality and identification of honey varieties for many years [19–21].

This study focused on refining the method and measurement conditions, as well as statistical tools for data analysis. The assumed effect of this study is a methodology, the application of which is to be a universal tool used on a wider scale. Creating fluorescence spectral matrices for honeys of various botanical origins will streamline the identification process, reduce the measurement time, and above all reduce the amount of reagents used and the costs incurred. Additionally, these tests are non-destructive with a high level of precision, sensitivity and repeatability of measurement performance. The preparation of the sample for testing does not require major expenditures. Therefore, the methodology can also be implemented to assess the quality of other parameters, such as the degree of filtration, honey aging, its overheating or its storage in improper conditions [22–24].

Therefore, the aim of our work was to demonstrate that fluorescence spectrometry can be such a method. Fluorescence is one of the physical techniques that is widely used for food authentication. Fluorescence spectra and images may both be considered unique sample fingerprints. This analytical technique is non-destructive, rapid and sensitive, especially when combined with multivariate analysis tools such as principal component analysis, parallel factor analysis and factorial discriminant analysis [25,26]. Numerous studies indicate that it can be used to mark the authenticity of olive oils, other edible oils, wines, spices, fish products and other foods [27–32]. In recent years, there have been many reports about using spectroscopic techniques to analyze honey as an alternative to time-consuming conventional methods. Fluorescence spectroscopy was used, among others, to determine the quality and authenticity of honey, but also to detect various contaminants or adulterants [33–39]. However, many of these reports show that determining the botanical origin of honey based only on fluorescence spectra is impossible. Therefore, we decided to demonstrate that such identification is possible both on the basis of 3D and synchronous spectra, but above all on the basis of differences in the fluorescence intensity of honeys of different varieties in the entire emission band, at a specified $\Delta\lambda = 100$ nm.

2. Materials and Methods

In total, 108 honey samples of 10 different botanical origins (multifloral: 14, honeydew: 12, acacia: 8, honeydew coniferous: 8, nectar honeydew: 8, rape: 13, phacelia: 9, lime: 14, buckwheat: 13, dandelion: 10, heather: 9) were analyzed to evaluate their fluorescence spectra. The honeys, provided by a local beekeepers' association from the Pomeranian province, were harvested in 2020–2022. Until the analyses were performed, the tested samples were stored in tightly closed individual packages—250 mL glass jars with a "Twist off" closure, at a temperature of 16 to 20 °C.

Fluorescence spectra were determined by a method patented by Gębala and Przybyłowski [40,41].

Excitation–emission matrices, EEMs, were acquired by a series of emission scans measured over a range of excitation wavelengths to create a fluorescence contour map.

The tests were performed using a set consisting of the following:

1. A Hitachi F-7000 fluorescence spectrophotometer, Tokyo, Japan;
2. A specially constructed adapter for measuring fluorescence intensity measured from the sample surface to change its traditional measurement range.

Figure 1 shows how the energy of the light beam reappears and re-emits at lower quanta after being reflected from the surface of the tested sample.

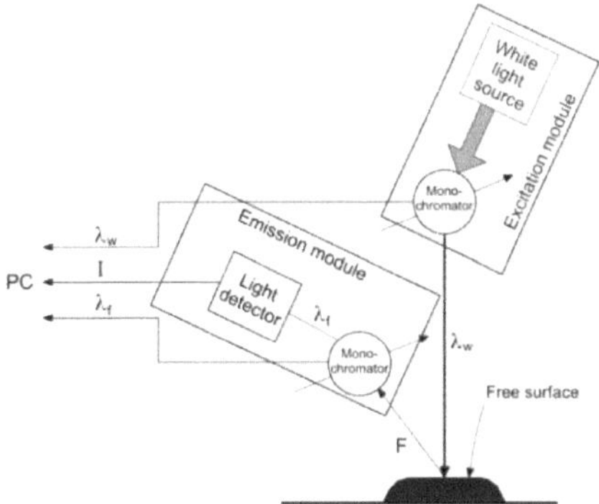

Figure 1. Scheme of surface fluorescence measurement. Source: (adapted from: [40,41]).

The sample surface is exposed to excitation radiation and the reflective geometry allows us to eliminate the effects of the internal filter associated with the high absorbance of the sample—the weakening of fluorescence intensity is due to the excitation absorbance and the emitted radial radiation.

Dimensional fluorescence spectra were measured at room temperature and in daylight. Honey samples were condensed (temp. 40 °C) and pipetted into 0.5 mL quartz cuvettes before measurement. Fluorescence spectra were obtained by recording emission spectra (from 220 to 560 nm with 10 nm steps) corresponding to excitation wavelengths in the range from 200 to 450 nm (with 5 nm steps) and automatically normalized to the excitation intensity of the instrument. The voltage used to determine the sensitivity of the excitation and emission measurements was equal to 600 V. The difference between the wavelength of fluorescent light (γF) and the wavelength of excitation light (γw) was preferentially 100 nm [40,41]. In order to reduce scattering effects and compare the honey samples studied, the fluorescence spectra were normalized by reducing the area under each spectrum to the value of 1 [22].

All analyses were performed in triplicate. The final results are presented as a set of numerical data in the form of contour maps (excitation emission (EEM)) and the synchronous cross-sections of these spectra were obtained at $\Delta\lambda = 100$ nm for honey samples.

Statistical Analysis

To determine the botanical origin on the shape of the spectra, a discriminant analysis was performed. The analysis was performed using the Statistica 13.3 multivariate discriminant analysis package.

This article uses discriminant analysis, which is a set of statistical methods for multi-dimensional data analysis. Its main goal is to decide which independent variables affect the classification of the described dependent variables. The classical discriminant analysis used in this study allows us to build a forecast model of group membership. This model is created on the basis of a discriminant function that provides the best distinction between groups. The sample of observations used to generate the function is known. It is an extremely effective tool used for classification issues and data exploration. Its advantage is a high level of effectiveness for homogeneous data, while its disadvantage is a lack of effectiveness on non-homogeneous data.

This analysis was aimed at confirming the hypothesis that the identification of honey varieties is possible based on fluorescence spectra. It was therefore checked whether the compared varieties differed in terms of the value of the mean fluorescence intensity within the entire emission band or in part of it at $\Delta\lambda$ 60 nm, 80 nm and 100 nm. However, attempts to perform calculations at $\Delta\lambda$ 60 nm and 80 nm did not produce results that provided statistically significant information. Then, appropriate canonical discriminant functions were constructed. Statistical hypotheses were verified at a significance level of $p = 0.05$.

3. Results

3.1. Analysis of Spectra

The emission spectra (excitation: 200–450 nm; emission: 260–560 nm) considered in this investigation allowed for the study of the fluorescence of the honey samples. The complete fluorescence spectra of different botanical origins of honey are characterized by a spectral region of high emission intensity, originating from fluorophores such as phenolic compounds and aromatic amino acids [26].

Figure 2 shows the EEM spectra measured for the tested honeys. The results show contour maps for 108 samples of different honey types: multifloral, acacia, honeydew coniferous, nectar honeydew, rape, phacelia, lime, buckwheat, dandelion and heather. The fluorescence measurement of the sample surface was performed three times.

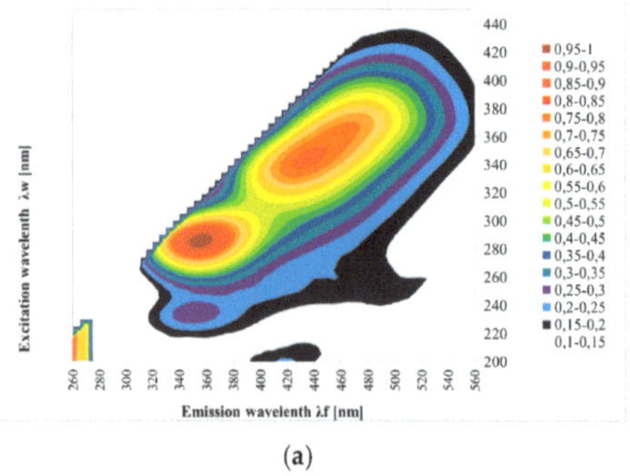
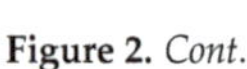

(a)

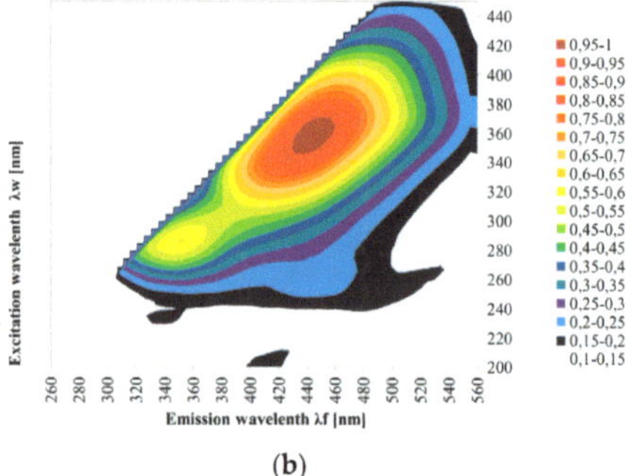

(b)

Figure 2. *Cont.*

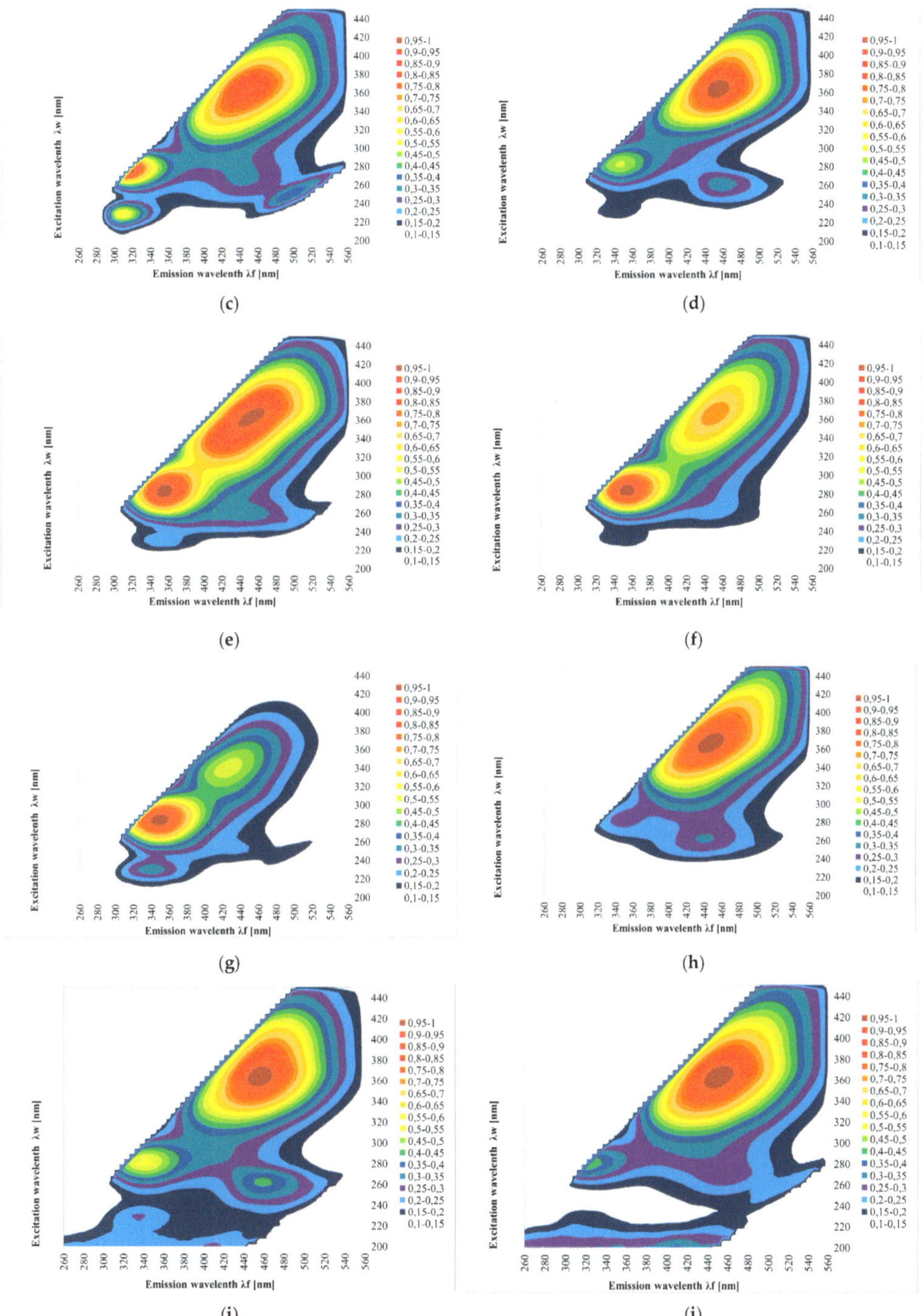

Figure 2. Excitation emission (EEM) spectra of different botanical origins of honey; (**a**)—acacia, (**b**)—phacelia, (**c**)—buckwheat, (**d**)—linden, (**e**)—dandelion, (**f**)—nectar–honeydew, (**g**)—rapeseed, (**h**)—honeydew–coniferous, (**i**)—multifloral, (**j**)—heather. Source: own research.

3.2. Discrimination of Honeys by Mathematical Method

In the next part, a discriminant analysis was performed to confirm the hypothesis that the identification of honey varieties is possible based on fluorescence spectra. It was therefore checked whether the compared varieties differ in terms of the value of the average fluorescence intensity within the entire emission band or in part of it at $\Delta\lambda$ 100 nm. Then, appropriate canonical discriminant functions were constructed.

Based on the analysis of the constructed classification matrices, that the following was found:

- The selection of any model covering only a part of the emission band does not allow us to obtain a 100% correct classification of all honey samples;
- After introducing all of the measurement points of the emission band into the model, a 100% qualification of varietal honeys was obtained at $\Delta\lambda$ 100 nm within the entire emission band (Table 1).

Table 1. Honey variety classification matrix at $\Delta\lambda$ 100 nm in the entire emission band.

Honey	% Correct Classification	Acacia	Phacelia	Buckwheat	Linden	Dandelion	Nectar–Honeydew	Rapeseed	Honeydew–Confireous	Multifloral	Heather
acacia	100	6	0	0	0	0	0	0	0	0	0
phacelia	100	0	6	0	0	0	0	0	0	0	0
buckwheat	100	0	0	8	0	0	0	0	0	0	0
linden	100	0	0	0	12	0	0	0	0	0	0
dandelion	100	0	0	0	0	6	0	0	0	0	0
Nectar–honeydew	100	0	0	0	0	0	5	0	0	0	0
rapeseed	100	0	0	0	0	0	0	10	0	0	0
Honeydew–needle	100	0	0	0	0	0	0	0	6	0	0
multifloral	100	0	0	0	0	0	0	0	0	14	0
heather	100	0	0	0	0	0	0	0	0	0	8

Source: own research.

The parameters of the measurements of the excitation waves of honeys for $\Delta\lambda = 100$ nm of different botanical origins were different. This allowed for the discrimination of the studied population of honeys by variety, which was statistically confirmed by the Wilks lambda test (F = 3.91, $p = 0.0000001$), assuming a test probability value of $p \leq 0.05$. This means that based on the measurement of the fluorescence intensity of honeys, their botanical identification was achieved in the case of 80.0% of the tested samples (the identification of dandelion honey may be difficult because it is located in the center of the coordinate system and close to three other varieties).

The main stage of the discriminant analysis was performed using the standard method. The canonical analysis performed next allowed for the calculation of raw coefficient values (F_1 and F_2) of the discriminant function for the first two roots and the construction of canonical discriminant functions.

The canonical analysis included also the calculation of the coefficients of canonical variables (D_1 and D_2) and their average values (Table 2).

Table 2. Analysis of fluorescence intensity in the entire emission band at $\Delta\lambda = 100$ nm—averages of canonical variables.

Canonical Variables	Honey									
	Acacia	Phacelia	Buckwheat	Linden	Dandelion	Nectar–Honeydew	Rapeseed	Honeydew–Confireous	Multifloral	Heather
	Averages of Canonical Variates									
D_1	−21.71	−16.19	4.434	−3.644	16.38	6.091	−0.367	−4.99	−4.901	26.144
D_2	14.3	3.88	2.54	2.564	−23.47	−8.336	−0.778	−10.86	−4.607	19.975

Source: own research.

The canonical analysis performed allowed for a graphical presentation of the results of the calculations of the canonical mean values (Figure 3). It illustrates the position of individual honey varieties in a two-coordinate system (canonical variables).

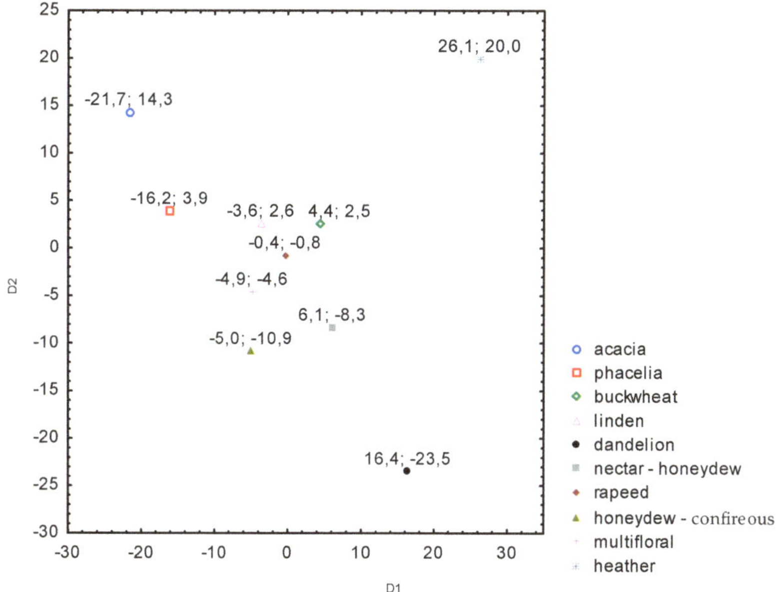

Figure 3. Analysis of the fluorescence intensity of honeys of different varieties in the entire emission band at $\Delta\lambda = 100$ nm—canonical variate averages. Source: own research.

Table 3 shows the outline of the spreadsheet file that can be used to calculate the values of two canonical variates—D_1 and D_2.

Using the above model, it is possible to identify all ten honey varieties using fluorescence spectra for $\Delta = 100$ nm. To do this, you should do the following:

- Calculate the D_1 and D_2 values according to the scheme presented in Table 3;
- Compare their values with the values given in Table 2, which contains the values of the mean canonical variables.

Table 3. Analysis of the fluorescence intensity of honeys of different varieties in the entire emission band at $\Delta\lambda = 100$ nm—scheme for calculating the value D_1 and D_2.

Emission Band	Fluorescence Intensity	F_1 (Discriminant Function Coefficients)	Square 1× Fluorescence Intensity	F_2 (Discriminant Function Coefficients)	Square 2× Fluorescence Intensity
Free term	-	−3.664	−3.664	2.677	2.677
200.000		−0.502		0.079	
205.000		−0.717		−0.577	
210.000		1.398		0.927	
215.000		−1.146		−1.452	
220.000		1.239		1.299	
225.000		1.425		0.840	
230.000		0.824		−0.461	
235.000		−2.946		−2.929	
240.000		−0.666		4.322	
245.000		3.522		−1.546	
250.000		−0.464		0.131	
255.000		−1.920		−0.054	
260.000		−0.696		−2.729	
265.000		−2.201		2.347	
270.000		1.950		1.718	
275.000		0.763		−3.293	
280.000		2.151		1.787	
285.000		−1.074		0.707	
290.000		0.056		−0.231	
295.000		−3.589		−2.249	
300.000		1.211		−0.335	
305.000		2.889		3.227	
310.000		−1.094		−1.092	
315.000		−1.987		−2.966	
320.000		−0.798		0.206	
325.000		4.240		1.035	
330.000		−0.729		4.727	
335.000		−1.849		−3.363	
340.000		1.717		−1.704	
345.000		1.834		4.251	
350.000		−3.414		−2.002	
355.000		−4.152		−2.765	
360.000		4.684		1.588	
365.000		−0.456		0.161	
370.000		6.187		3.346	
375.000		−4.361		−5.826	
380.000		−4.968		2.121	
385.000		5.438		3.984	
390.000		−2.449		−3.884	
395.000		0.741		3.593	
400.000		−5.072		−4.049	
405.000		4.449		−0.149	
410.000		2.754		2.695	
415.000		−6.915		−0.923	
420.000		2.698		−1.150	
425.000		3.570		−0.825	
430.000		−1.298		2.171	
435.000		4.690		−4.005	
440.000		−1.523		0.744	
445.000		−8.798		−0.297	
450.000		5.980		3.072	
			TOTAL = D_1		TOTAL = D_2

Source: own research.

Based on the analysis carried out, taking into account the difficult identification and analysis of the spectra, it was found that the greatest essence of information concerning botanical differences and their identification is contained in the synchronous sections of these spectra obtained at $\Delta\lambda = 100$ nm, which are shown in Figures 4 and 5. It can also be seen that the spectra of all tested honeys in a given variety group were characterized by the presence of emission bands of different intensities, but the same locations of the excitation wavelength $\Delta\lambda w$ for the maximum level of fluorescence intensity. Figure 4 shows the tested honeys, taking into account each of the tested samples in the variety group. Each sample was marked with a different marker in order to show the deviations of the fluorescence measurements in the variety group.

The spectra of all the honeys tested, depending on their botanical origin, were characterized by the presence of individual emission bands of different intensities. Additionally, the bands were distinguished by different locations of the excitation wavelength $\Delta\lambda w$ for the maximum level of fluorescence intensity at $\Delta\lambda = 100$ nm (Figure 5). It can be stated that in eight varieties, there were three excitations at different excitation wavelengths $\Delta\lambda w$, which is also presented in more detail in Table 4. In the heather and honeydew–conifer varieties, there are two excitations at different excitation wavelengths $\Delta\lambda w$, 235 nm and 360 nm for honeydew–conifer honeys and 370 nm for heather honeys. All the honey varieties tested had a visible tendency to increase the intensity of the shortwave emission excitation band at the level of 235 nm and 240 nm (phacelia). Another increase in intensity in the intermediate-range bands was found in eight varieties:

- Emission excitation: 280 nm (honey: acacia, buckwheat, lime, rapeseed, multifloral);
- Emission excitation: 285 nm (honey: phacelia, dandelion, nectar–honeydew).

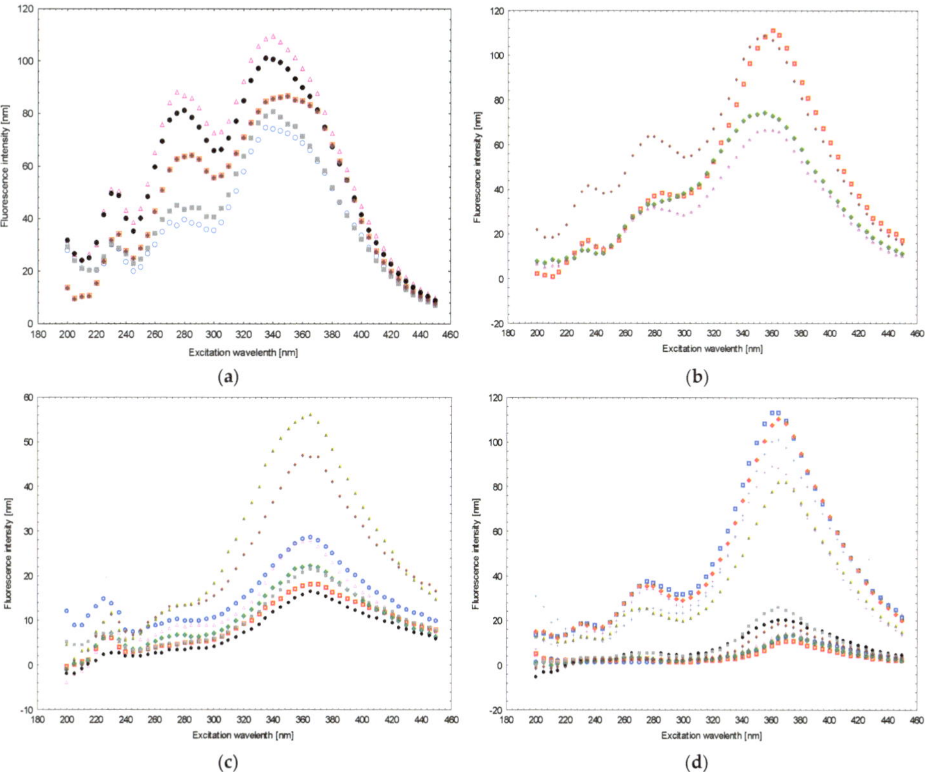

Figure 4. *Cont.*

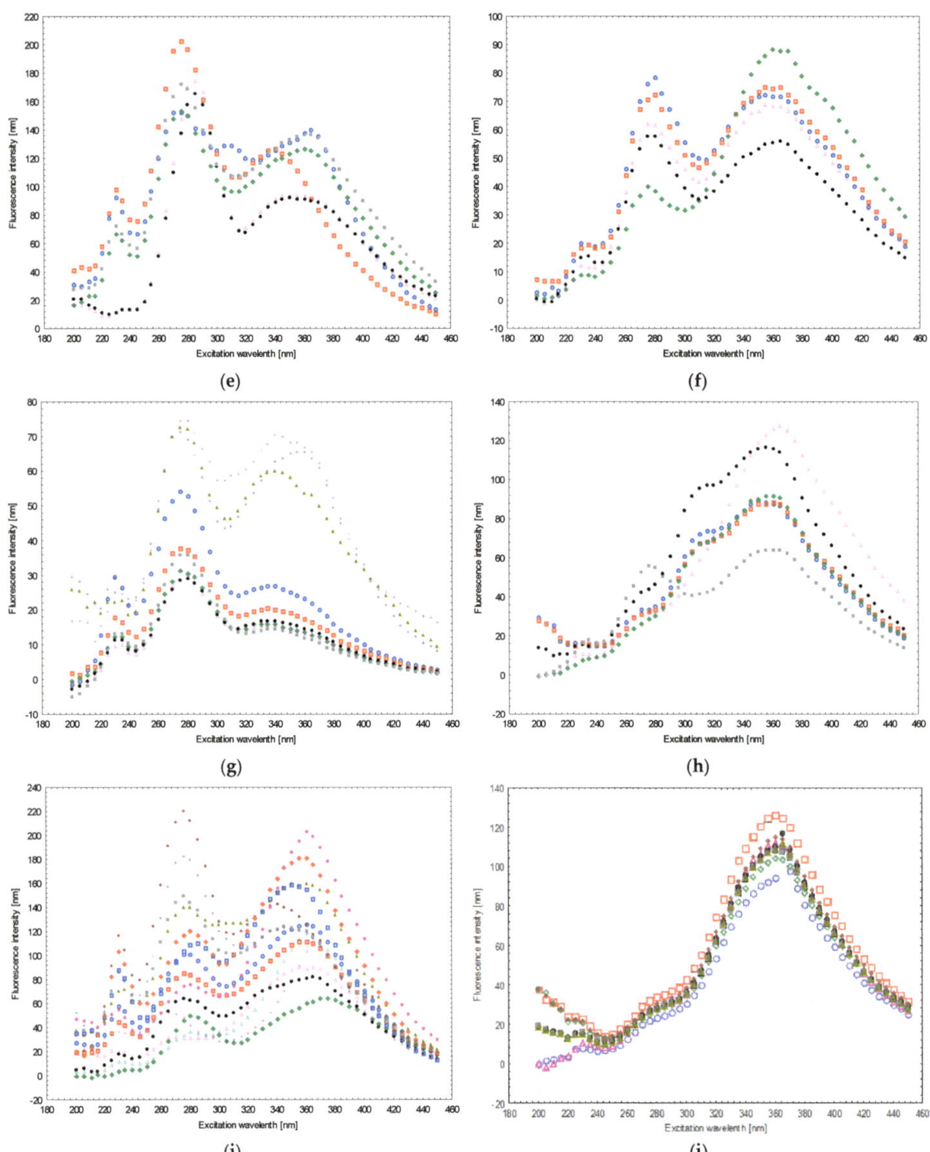

Figure 4. Synchronous fluorescence spectrum of honeys of different botanical origin at Δλ = 100 nm with each sample marked in the species group; (**a**)—acacia, (**b**)—phacelia, (**c**)—buckwheat, (**d**)—linden, (**e**)—dandelion, (**f**)—nectar–honeydew, (**g**)—rapeseed, (**h**)—honeydew–coniferous, (**i**)—multiflora, (**j**)—heather. Different colors shown in the figure indicate separate samples. Source: own research.

The next increase in intensity in the bands in the long-wave spectral range (excitation of emission from 340 nm to 370 nm occurred in all ten varieties tested).

In our analysis, the synchronous spectrum of buckwheat and lime honey (Δλ = 100 nm) was characterized by a visible tendency to demonstrate three emission excitations, 235 nm, 280 nm and 370 nm, with different fluorescence emission intensities.

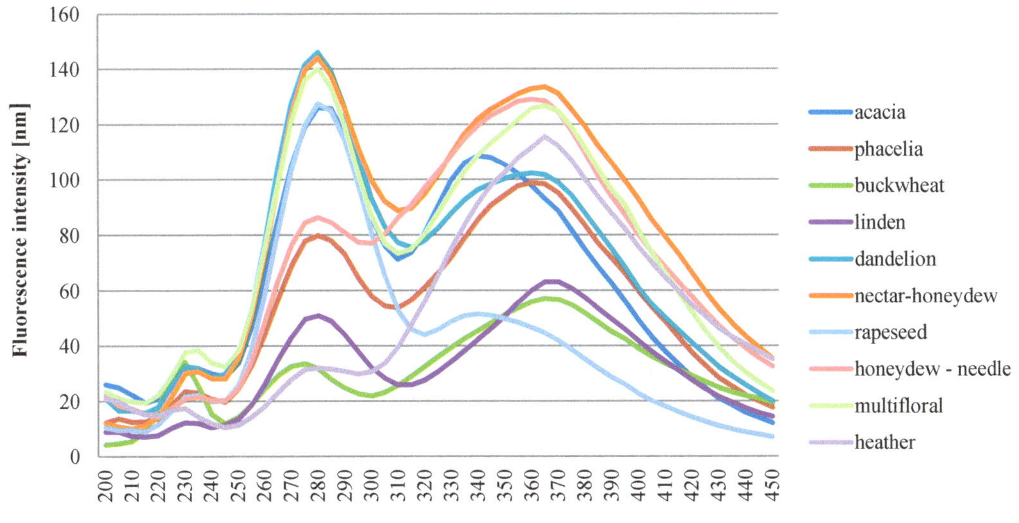

Figure 5. Synchronous fluorescence spectrum for varietal honeys at Δλ = 100 nm (mean value). Source: own research.

Table 4. Comparison of excitation wavelengths Δλw for the maximum fluorescence intensity of honeys of different botanical origin at Δλ = 100 nm—average.

Botanical Origin of Honey	Excitation Wavelength Δλw for Maximum Fluorescence Intensity of Varietal Honeys [nm]		
	I Excitation	**II Excitation**	**III Excitation**
acacia	235	280	340
phacelia	240	285	360
buckwheat	235	280	370
linden	235	280	370
dandelion	235	285	365
nectar–honeydew	235	285	360
rapeseed	235	280	340
honeydew–needle	235	absence	360
multifloral	235	280	365
heather	235	absence	370

Source: own research.

4. Discussion

It can be argued that the results of the use of fluorescence spectrometry in the form of total fluorescence spectra depend on the botanical origin of the honey, and this was confirmed by the studies presented by Ruoff et al. [19–21], Gębala [40], Gębala and Przybyłowski [41], Dramićanin et al. [39] and our former studies [22]. These authors also indicated small differences in fluorescence intensity in the tested group of varietal honeys. Gębala and Przybyłowski grouped the spectral features of Polish botanical honeys, taking into account the levels of excitation fluorescence intensity as indicators characteristic of varietal honeys found in Poland [40]. Pari et al. grouped samples of honeys of Italian origin into subsections of honeys of different botanical origins. The grouping was based on similar spectral features. The authors also point to the main components, the so-called fluorophores, which are responsible for the fluorescence emission of the honeys studied [33]. In 2014, Nikolova's team attempted to compare the spectral characteristics of honeys adulterated by the addition of sweeteners. The results obtained indicated that the study of fluorescence spectra opens up the possibility of distinguishing honey samples with added artificial sweeteners from natural ones [42]. In our previous studies, we attempted to demonstrate that fluorescence spectra can be used to indicate the degree of overheating of honey [23]. We also obtained confirmation that honey filtration does not affect the classification of honey spectra of various botanical origins [22]. According to the study conducted by Dramićanin et al. [39], the differences in the fluorescence of natural and adulterated honey samples are extremely significant in five spectral regions due to the differences in the concentrations and local environments of aromatic amino acids, phenolic compounds, furosine and hydroxymethylfurfural. This was also confirmed by statistical tests and PCA. According to these authors, by quantifying the fluorescence responses and subjecting them to a statistical classification technique, for example, LDA, it is possible to detect adulterated honey with 100% accuracy. Such accuracy suggests that fluorescence excitation emission spectroscopy may be a promising method for low-level honey adulteration, which will be the subject of our future work.

5. Conclusions

In our study, 108 pure honey samples from 10 different botanical origins have been analyzed using fluorescence emission. We have shown that each honey variety is characterized by a unique total spectrum, enabling the classification of their botanical origin. However, taking into account the difficulties in analyzing these spectra, it was found that the most information regarding botanical differences and their identification is provided by synchronous cross-sections of these spectra obtained at $\Delta\lambda = 100$ nm. Additionally, discriminant analysis showed that it was possible to identify most of the tested honey varieties; the identification of dandelion honey will be problematic.

Author Contributions: Conceptualization, A.W. and N.Ż.; methodology, A.W. and N.Ż.; formal analysis, A.W. and N.Ż.; data curation, N.Ż.; writing—original draft preparation, N.Ż. and A.W.; writing—review and editing, A.W.; supervision, A.W. All authors have read and agreed to the published version of the manuscript.

Funding: The authors acknowledge the financial support of Gdynia Maritime University (projects: WZNJ/2024/PZ/01, WZNJ/2024/PI/01).

Data Availability Statement: The original contributions presented in the study are included in the article, further inquiries can be directed to the corresponding author.

Conflicts of Interest: The authors declare no conflict of interest.

References

1. Directive2014/63/EU of the European Parliament and of the Council of 15 May 2014 amending Council Directive 2001/110/EC Relating to Honey. Available online: https://eur-lex.europa.eu/legal-content/EN/TXT/?uri=celex:32014L0063 (accessed on 14 October 2024).
2. Escuredo, O.; Rodríguez-Flores, M.S.; Rojo-Martínez, S.; Seijo, M.C. Contribution to the Chromatic Characterization of Unifloral Honeys from Galicia (NW Spain). *Foods* **2019**, *8*, 233. [CrossRef]
3. Bogdanov, S. Honey as Nutrient and Functional Food: A Review. Book of Honey (Chapter 8). Available online: https://freethebees.ch/wp-content/uploads/2020/06/HoneyNutrientFunctional.pdf (accessed on 14 October 2024).
4. Biswas, A.; Chaudchari, S. Exploring the role of NIR spectroscopy in quantifying and verifying honey authenticity: A review. *Food Chem.* **2024**, *445*, 138712. [CrossRef]
5. Quintanilla-López, J.E.; Lebrón-Aguilar, R.; Soria, A.C. Volatile fingerprinting by solid-phase microextraction mass spectrometry for rapid classification of honey botanical source. *LWT* **2022**, *169*, 114017. [CrossRef]
6. Biswas, A.; Naresh, K.S.; Jaygadkar, S.S.; Chaudhari, S.R. Enabling honey quality and authenticity with NMR and LC-IRMS based platform. *Food Chem.* **2023**, *416*, 135825–135838. [CrossRef]
7. Bodor, R.; Kovacs, Z.; Benedek, C.; Hitka, G.; Behling, H. Origin identification of hungarian honey using melissopalynology, physicochemical analysis, and near infrared spectroscopy. *Molecules* **2021**, *26*, 7274. [CrossRef]

8.   Da Silva, P.M.; Gauche, C.; Gonzaga, L.V.; Costa, A.C.O.; Fett, R. Honey: Chemical composition, stability and authenticity. *Food Chem.* **2016**, *196*, 309–323. [CrossRef]
9.   Dong, H.; Xiao, K.; Xian, Y.; Wu, Y. Authenticity determination of honeys with non-extractable proteins by means of elemental analyzer (EA) and liquid chromatography (LC) coupled to isotope ratio mass spectroscopy (IRMS). *Food Chem.* **2018**, *240*, 717–724. [CrossRef]
10. Gan, Z.; Yang, Y.; Li, J.; Wen, X.; Zhu, M.; Jiang, Y.; Ni, Y. Using sensor and spectral analysis to classify botanical origin and determine adulteration of raw honey. *J. Food Eng.* **2016**, *178*, 151–158. [CrossRef]
11. Machado, A.M.; Miguel, M.G.; Vilas-Boas, M.; Figueiredo, A.C. Honey Volatiles as a Fingerprint for Botanical Origin—A Review on their Occurrence on Monofloral Honeys. *Molecules* **2020**, *25*, 374. [CrossRef]
12. Żak, N.; Wilczyńska, A. The Importance of Testing the Quality and Authenticity of Food Products: The Example of Honey. *Foods* **2023**, *12*, 3210. [CrossRef]
13. Zhang, X.-H.; Gu, H.-W.; Liu, R.-J.; Qing, X.-D.; Nie, J.-F. A comprehensive review of the current trends and recent advancements on the authenticity of honey. *Food Chem.* **2023**, *19*, 100850. [CrossRef]
14. Tsagkaris, A.S.; Koulis, G.A.; Danezis, G.P.; Martakos, I.; Dasenaki, M.; Georgiou, C.A.; Thomaidis, N.S. Honey authenticity: Analytical techniques, state of the art and challenges. *RSC Adv.* **2021**, *11*, 11273–11294. [CrossRef]
15. Guzelmeric, E.; Ciftci, I.; Yuksel, P.I.; Yesilada, E. Importance of chromatographic and spectrophotometric methods in determining authenticity, classification and bioactivity of honey. *LWT* **2020**, *132*, 109921. [CrossRef]
16. Antônio, D.C.; de Assis, D.C.S.; Botelho, B.G.; Sena, M.M. Detection of adulterations in a valuable Brazilian honey by using spectrofluorimetry and multiway classification. *Food Chem.* **2022**, *370*, 131064. [CrossRef]
17. Rahman, M.M.; Bui, M.V.; Shibata, M.; Nakazawa, N.; Rithu, M.N.A.; Yamashita, H.; Sadayasu, K.; Tsuchiyama, K.; Nakauchi, S.; Hagiwara, T.; et al. Rapid noninvasive monitoring of freshness variation in frozen shrimp using multidimensional fluorescence imaging coupled with chemometrics. *Talanta* **2021**, *224*, 121871. [CrossRef]
18. Gu, H.; Huang, X.; Chen, Q.; Sun, Y.; Lv, R. A feasibility study for rapid evaluation of emulsion oxidation using synchronous fluorescence spectroscopy coupled with chemometrics. *Spectrochim. Acta Part A Mol. Biomol. Spectrosc.* **2022**, *265*, 120337. [CrossRef]
19. Ruoff, K.; Iglesias, M.T.; Luginbühl, W.; Bosset, J.O.; Bogdanov, S.; Amado, R. Quantitative analysis of physical and chemical measurands in honey by mid-infrared spectrometry. *Eur. Food Res. Technol.* **2005**, *223*, 22–29. [CrossRef]
20. Ruoff, K.; Karoui, R.; Dufour, E.; Luginbühl, W.; Bosset, J.O.; Bogdanov, S.; Amadò, R. Authentication of the botanical origin of honey by front-face fluorescence spectroscopy, a preliminary study. *J. Agric. Food Chem.* **2005**, *53*, 1343–1347. [CrossRef]
21. Ruoff, K.; Luginbühl, W.; Künzli, R.; Bogdanov, S.; Bosset, J.O.; von der Ohe, K.; von der Ohe, W.; Amadò, R. Authentication of the botanical and geographical origin of honey by front-face fluorescence spectroscopy. *J. Agric. Food Chem.* **2006**, *54*, 6858–6866. [CrossRef]
22. Wilczyńska, A.; Żak, N. The Use of Fluorescence Spectrometry to Determine the Botanical Origin of Filtered Honeys. *Molecules* **2020**, *25*, 1350. [CrossRef]
23. Żak, N.; Wilczyńska, A.; Przybyłowski, P. The use of fluorescence spectroscopy to assess the degree of overheating of honey. *Folia Pomer. Univ. Technol. Stetin. Agric. Aliment. Pisc. Zootech.* **2018**, *340*, 131–142. [CrossRef]
24. Gu, H.; Hu, L.; Dong, Y.; Chen, Q.; Wei, Z.; Lv, R. Evolving trends in fluorescence spectroscopy techniques for food quality and safety: A review. *J. Food Comp. Anal.* **2024**, *131*, 106212. [CrossRef]
25. Dankowska, A. Advances in Fluorescence Emission Spectroscopy for Food Authenticity Testing. In *Woodhead Publishing Series in Food Science, Technology and Nutrition, Advances in Food Authenticity Testing*; Gerard, D., Ed.; Woodhead Publishing: Sawston, UK, 2016; pp. 117–145. [CrossRef]
26. Sikorska, E.; Khmelinskii, I.; Sikorski, M. Fluorescence spectroscopy and imaging instruments for food quality evaluation. In *Woodhead Publishing Series in Food Science, Technology and Nutrition, Evaluation Technologies for Food Quality*; Jian, Z., Xichang, W., Eds.; Woodhead Publishing: Sawston, UK, 2019; pp. 491–533. [CrossRef]
27. Hassoun, A.; Sahar, A.; Lakhal, L.; Aït-Kaddour, A. Fluorescence spectroscopy as a rapid and non-destructive method for monitoring quality and authenticity of fish and meat products: Impact of different preservation conditions. *LWT* **2019**, *103*, 279–292. [CrossRef]
28. Wu, Q.; Geng, T.; Yan, M.-L.; Peng, Z.-X.; Chen, Y.; Lv, Y.; Yin, X.-L.; Gu, H.-W. Geographical origin traceability and authenticity detection of Chinese red wines based on excitation-emission matrix fluorescence spectroscopy and chemometric methods. *J. Food Comp. Anal.* **2024**, *125*, 105763. [CrossRef]
29. Durán Merás, I.; Domínguez Manzano, J.; Airado Rodríguez, D.; Muñoz de la Peña, A. Detection and quantification of extra virgin olive oil adulteration by means of autofluorescence excitation-emission profiles combined with multi-way classification. *Talanta* **2018**, *178*, 751–762. [CrossRef]
30. Ríos-Reina, R.; Ocaña, J.A.; Azcarate, S.M.; Pérez-Bernal, J.L.; Villar-Navarro, M.; Callejón, R.M. Excitation-emission fluorescence as a tool to assess the presence of grape-must caramel in PDO wine vinegars. *Food Chem.* **2019**, *287*, 115–125. [CrossRef]
31. Fang, H.; Wang, T.; Chen, L.; Wang, X.-Z.; Wu, H.-L.; Chen, Y.; Yu, R.-Q. Rapid authenticity identification of high-quality Wuyi Rock tea by multidimensional fluorescence spectroscopy coupled with chemometrics. *J. Food Comp. Anal.* **2024**, *135*, 106632. [CrossRef]

32. Yuan, Y.-Y.; Wang, S.-T.; Wang, J.-Z.; Cheng, Q.; Wu, X.-J.; Kong, D.-M. Rapid detection of the authenticity and adulteration of sesame oil using excitation-emission matrix fluorescence and chemometric methods. *Food Contr.* **2020**, *112*, 107145. [CrossRef]
33. Nikolova, K.; Eftimov, T.; Aladjadjiyan, A. Fluorescence Spectroscopy as Method for Quality Control of Honey. *Adv. Res.* **2014**, *2*, 95–108. [CrossRef]
34. Hao, S.; Li, J.; Liu, X.; Yuan, J.; Yuan, W.; Tian, Y.; Xuan, H. Authentication of acacia honey using fluorescence spectroscopy. *Food Contr.* **2021**, *130*, 108327. [CrossRef]
35. Chen, Q.; Qi, S.; Li, H.; Han, X.; Ouyang, Q.; Zhao, Q. Determination of rice syrup adulterant concentration in honey using three-dimensional fluorescence spectra and multivariate calibrations. *Spectrochim. Acta Part A Mol. Biomol. Spectrosc.* **2014**, *131*, 177–182. [CrossRef]
36. Santana, J.E.G.; Coutinho, H.D.M.; da Costa, J.G.M.; Menezes, J.M.C.; Pereira Teixeira, R.N. Fluorescent characteristics of bee honey constituents: A brief review. *Food Chem.* **2021**, *362*, 130174. [CrossRef]
37. Lastra-Mejías, M.; Torreblanca-Zanca, A.; Aroca-Santos, R.; Cancilla, J.C.; Izquierdo, J.G.; Torrecilla, J.S. Characterization of an array of honeys of different types and botanical origins through fluorescence emission based on LEDs. *Talanta* **2018**, *185*, 196–202. [CrossRef]
38. Sergiel, I.; Pohl, P.; Biesaga, M.; Mironczyk, A. Suitability of three-dimensional synchronous fluorescence spectroscopy for fingerprint analysis of honey samples with reference to their phenolic profiles. *Food Chem.* **2014**, *145*, 319–326. [CrossRef]
39. Dramićanin, T.; Lenhardt Acković, L.; Zeković, I.; Dramićanin, M.D. Detection of Adulterated Honey by Fluorescence Excitation-Emission Matrices. *J. Spectrosc.* **2018**, *2018*, 8395212. [CrossRef]
40. Gębala, S. Measurements of solution fluorescence–A new concept. *Opt. Appl.* **2009**, *39*, 391–399.
41. Gębala, S.; Przybyłowski, P. Sposób Identyfikacji Odmian Miodu. Polska Patent 214784, 15 March 2010.
42. Parri, E.; Santinami, G.; Domenici, V. Front-Face Fluorescence of Honey of Different Botanic Origin: A Case Study from Tuscany (Italy). *Appl. Sci.* **2020**, *10*, 1776. [CrossRef]

Communication

(−)-Gallocatechin Gallate: A Novel Chemical Marker to Distinguish *Triadica cochinchinensis* Honey

Huizhi Jiang [1,2,†], Zhen Li [3,†], Shiqing Zhong [1,2] and Zhijiang Zeng [1,2,*]

1. Honeybee Research Institute, Jiangxi Agricultural University, Nanchang 330045, China; jhz19991221@163.com (H.J.); zhongsqing@yeah.net (S.Z.)
2. Jiangxi Province Key Laboratory of Honeybee Biology and Beekeeping, Jiangxi Agricultural University, Nanchang 330045, China
3. College of Life Science and Resources and Environment, Yichun University, Yichun 336000, China; zhenli1995@aliyun.com
* Correspondence: bees@jxau.edu.cn; Tel.: +86-791-8381-3998
† These authors contributed equally to this work.

Abstract: *Triadica cochinchinensis* honey (TCH) is collected from the nectar of the medicinal plant *T. cochinchinensis* and is considered the most important honey variety in southern China. TCH has significant potential medicinal properties and commercial value. However, reliable markers for application in the authentication of TCH have not yet been established. Herein, a comprehensive characterization of the botanical origin and composition of TCH was conducted by determining the palynological characteristics and basic physicochemical parameters. Liquid chromatography tandem-mass spectrometry (LC-MS/MS) was used to investigate the flavonoid profile composition of TCH, *T. cochinchinensis* nectar (TCN) and 11 other common varieties of Chinese commercial honey. (−)-Gallocatechin gallate (GCG) was identified as a reliable flavonoid marker for TCH, which was uniquely shared with TCN but absent in the other 11 honey types. Furthermore, the authentication method was validated, and an accurate quantification of GCG in TCH and TCN was conducted. Overall, GCG can be applied as a characteristic marker to identify the botanical origin of TCH.

Keywords: *Triadica cochinchinensis*; honey; LC-MS/MS; (−)-gallocatechin gallate; chemical markers

Citation: Jiang, H.; Li, Z.; Zhong, S.; Zeng, Z. (−)-Gallocatechin Gallate: A Novel Chemical Marker to Distinguish *Triadica cochinchinensis* Honey. *Foods* **2024**, *13*, 1879. https://doi.org/10.3390/foods13121879

Academic Editor: Olga Escuredo

Received: 8 May 2024
Revised: 8 June 2024
Accepted: 13 June 2024
Published: 14 June 2024

1. Introduction

Honey is a natural sweet substance produced by honeybees (*Apis mellifera*) and derived from the nectar or secretions of plants or the excretions of plant-sucking insects on the living parts of plants [1]. It is known that the chemical composition and biological activities of honey are mainly influenced by nectar source plants [2].

Honey contains approximately 20% water and 75% carbohydrates (mainly fructose and glucose), as well as flavor components, proteins, minerals and phenolic compounds. These trace components contribute candidate markers tracing the botanical origin of honey [3,4]. Among the minor components, phenolic compounds are often paid more attention due to their contribution to the unique chemical profile of honey and used as chemical markers for distinguishing botanical origins [5]. Commonly, phenolic compounds can be divided into phenolic acids and flavonoids. According to previous reports, flavonoids are the most important phenolic constituent in honey (accounting for more than 80%) [6].

Honey is broadly classified based on its botanical origin into monofloral and polyfloral types [7]. Monofloral honey is widely considered more valuable than polyfloral honey owing to its distinct flavor and pharmacological properties and, in recent years, has witnessed increased consumer demand [8,9]. It has been suggested that several medicinal properties of plants can be carried on to monofloral honey [10], and recent studies have focused on identifying chemical markers that would enable the authentication of high-quality monofloral honey originating from medicinal plants. For instance, Manuka honey is renowned for its antibacterial properties, which can be identified by chemical markers such as methyl syringate 4-O-β-D-gentiobioside and lepteridine [11,12]. Moreover, kaempferitrin is a unique flavonoid that can be used as a marker to authenticate honey obtained from the nectar of the medicinal plant *Camellia oleifera* [13]. Furthermore, safflomin A is a novel chemical marker used for the authentication of honey derived from *Carthamus tinctorius* L. (Safflower), a well-known medicinal plant belonging to the Asteraceae family [14]. In addition, kaempferol-3-O-galactoside has been proposed as a marker for authenticating honey from *Lespedeza bicolor* Turcz., which has highly valuable and relatively rare medicinal properties. Moreover, calycosin and formononetin have emerged as markers for honey from *Astragalus membranaceus* var. *mongholicus* Hsiao [15].

Triadica cochinchinensis is a tree or shrub belonging to the Euphorbiaceae family (Figure 1A) and one of the main nectar plants in southern China during summer time, showing a long flowering stage and high nectar production [16,17]. Traditionally used in Chinese herbal medicine, its leaves, bark and roots can be used to treat various internal pathological conditions, including nephritis, oedema, ascites, constipation and dysuria, as well as external pathologies, such as allergic dermatitis, mastitis, bruises, wounds and snakebites [18,19]. Recent studies have shown that *T. cochinchinensis* leaf and stem parts contain various bioactive compounds, including diterpenoids, phenolics, flavonoids and tannins, which are the main components imparting antioxidant, anti-inflammatory, antibacterial, hepatoprotective and antidiabetic effects [20,21].

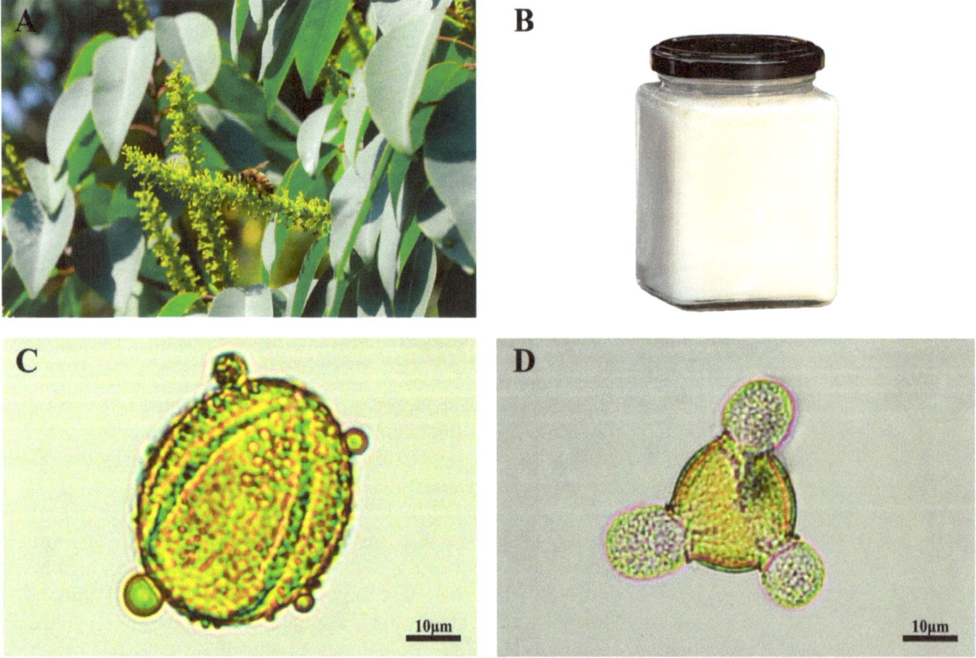

Figure 1. The botanical origin and palynological analysis of TCH. (**A**) Honeybee-visited *T. cochinchinensis* flower. (**B**) Crystallized *T. cochinchinensis* monofloral honey. (**C**) An equatorial view of TCH pollen grains under the light microscope. (**D**) A polar view of TCH pollen grains under the light microscope.

T. cochinchinensis honey (TCH) easily crystallizes, and crystallized TCH assumes a white color (Figure 1B). Previous studies have shown that TCH has demonstrated the ability of alleviating alcoholic liver damage and scavenging free radicals [21,22]. It has been described that TCH is rich in phenolic acids and flavonoids such as ellagic acid, gallic acid, naringenin and rutin [22,23]. Thus, TCH holds promise for the development of dietary supplements and medicinal agents.

Being a major commercial honey in southern China, *T. cochinchinensis* nectar (TCN) production is substantial and stable, making it one of the key monofloral honeys with significant economic benefits for beekeepers [24,25]. Therefore, identifying characteristic chemical markers in TCH derived from medicinal plants is crucial. Such an identification process might facilitate exploring efficacious and potential applications of TCH, thus contributing to further popularizing, while also enhancing, both the health benefits and commercial value of this unique monofloral honey. In addition, the identification of chemical markers would contribute in the evaluation of honey authenticity, which would help to promote consumer trust and sustain the reliable development of the food industry [26].

Previous studies have shown that chromatography and mass spectrometry are essential in identifying chemical markers specific to monofloral honey [27,28]. In particular, plant-nectar-derived flavonoids found in honey are additionally considered bioactive compounds and might be used as biomarkers for discriminating honey botanical origin as well as adulteration [29]. Several studies have proposed the use of flavonoids as unique chemical markers for specific monofloral honey [13,30]. Thus, utilizing distinct flavonoid markers might be considered a reliable strategy for authenticating TCH.

The present study aimed to identify distinctive flavonoid markers in TCH using targeted metabolomics. A palynological analysis was applied to confirm the botanical origin of TCH, and basic physicochemical parameters of TCH were analyzed. Subsequently, liquid chromatography tandem-mass spectrometry (LC-MS/MS) was used to identify flavonoid types in TCH, and a comparative analysis was conducted against 11 other common types of Chinese commercial honey to discriminate unique flavonoid markers exclusive to TCH. Finally, a method for TCH authentication and accurate quantification of flavonoid markers in TCH and TCN was proposed. This study will be helpful for authenticity assessment and quality control of TCH products.

2. Materials and Methods

2.1. Chemicals and Reagents

LC-MS-grade acetonitrile and methanol were purchased from Merck (Darmstadt, Germany). LC-MS-grade formic acid was purchased from Sigma-Aldrich (St. Louis, MO, USA). Ultra-pure water was obtained from a Milli-Q Plus system (Millipore, Bradford, PA, USA). All remaining standards (purity > 98%) were purchased from MedChemExpress (Shanghai, China). Detailed information on standards used in the present study is listed in Table S1.

2.2. Sample Collection

We selected the Weimin honeybee apiary in Yongxiu County, Jiujiang City, Jiangxi Province (29°01′ N, 115°29′ E) with *T. cochinchinensis* plantation areas to produce and harvest. TCH was collected from 28 May to 10 July 2023 when the flowers were blooming. TCN was collected from *T. cochinchinensis* plants by using a micro aspirator (Beijing Dalong Company Limited, Beijing, China).

Eleven types of popular commercial honey in China were selected as comparison honey, including nine types of monofloral honey, namely, *Brassica napus* honey (BNH), *Citrus reticulata* honey (CRH), *Robinia pseudoacacia* honey (RPH), *Ziziphus jujuba* honey (ZJH), *Vitex negundo* honey (VNH), *Litchi chinensis* honey (LCH), *Lycium chinense* honey (YCH), *Eriobotrya japonica* honey (EJH), *Tilia tuan* honey (TTH) and one polyfloral honey (POH) which were obtained from Wuhan Baochun Bee Products Company (Wuhan, China). *C. oleifera* honey (COH) was provided by Lishui Lantian Apiary in Changning City, China. All samples were prepared as three biological replicates and stored at −20 °C until analysis.

2.3. Sample Preparation

TCH and TCN samples were freeze-dried and ground into powder using a ball mill (MM400, Retsch, Haan, Germany) (30 Hz, 1.5 min). Then, 0.20 ± 0.01 g of samples was accurately weighted and mixed with 5 mL of 70% methanol and 100 µL of the internal standard working solution (daidzein, rutin and ($-$)-gallocatechin) of 4000 nmol/L. After ultrasonication for 30 min, samples were centrifuged at 12,000 r/min (the corresponding g-force was $11,304 \times g$) for 5 min at 4 °C. The obtained supernatant was harvested and filtered through a 0.22 µm filter membrane into a glass vial for subsequent LC-MS/MS analysis.

2.4. Palynological Identification and Physicochemical Analysis

Palynological analysis was conducted with a light microscope equipped with a camera (DS-Fi3, Nikon Corporation, Tokyo, Japan). Honey and pollen grain samples were prepared based on the method of Song et al. [31,32]. The concentration of *T. cochinchinensis* plant pollen grains in TCH pollen grains was determined. The morphology, length and width of *T. cochinchinensis* plants and TCH pollen grains were measured. The native pollen rate and morphological characterization of the other 10 types of monofloral honey samples were also determined with reference to the previous studies [33]. The contents of fructose, glucose, sucrose, water, 5-hydroxymethylfurfural (5-HMF), as well as diastase activity, electrical conductivity, ash content and color value in TCH samples were determined in accordance with the AOAC official method [34]. Additionally, free acidity in TCH was determined based on the equivalence point titration following the method of Li [13]. The concentration of pollen grains was calculated using a hemocytometer, and the counting method used for honey samples and pollen grains was conducted as proposed by Song [31]. Minerals Fe, Cu and Zn were identified using inductively coupled plasma optical emission spectrometry (ICP-MS 7500, Agilent Technologies Inc., Santa Clara, CA, USA) [35].

2.5. LC-MS/MS Analysis

For quantitative and qualitative analysis of flavonoid compounds in honey and nectar samples, an ultra-performance liquid chromatography system (ExionLC™ AD, SCIEX, Framingham, MA, USA) coupled with tandem mass spectrometry (QTRAP® 6500+, SCIEX, Framingham, MA, USA) was used. UPLC conditions were as follows: Waters ACQUITY UPLC HSS T3 C18 column (1.8 µm, 100 mm × 2.1 mm, Waters, Milford, MA, USA); mobile phase A, ultrapure water with 0.05% (v/v) formic acid; mobile phase B, acetonitrile with 0.05% (v/v) formic acid; flow rate, 0.35 mL/min; column temperature, 40 °C; injection volume, 2 µL. Gradient elution program was as follows: 0.0~1.0 min, 10~20% B; 1.0~9.0 min, 20~70% B; 9.0~12.5 min, 70~95% B; 12.5~13.5 min, 95%B; 13.5~13.6 min, 95~10% B; 13.6~15 min, 10% B.

Mass spectrometry (MS) conditions were as follows: electrospray ionization (ESI) source temperature, 550 °C; voltage in positive ion mode, 5500 V; voltage in negative ion mode, -4500 V; curtain gas (nitrogen), 35 psi. The collision gas was nitrogen. Each ion pair was scanned for detection in a Q-Trap 6500+ system (SCIEX, Framingham, MA, USA) on the basis of the optimized declustering potential (DP) and collision energy (CE). The specific flavonoid standards monitored in positive and negative ion modes are listed in Supplementary Table S2.

2.6. Construction of Flavonoid Standard Curves and Determination of Linearity Range

All 204 flavonoid standards were prepared into master batches of 10 mmol/L by methanol/water (70:30), then diluted in methanol/water (70:30) into standard curve working solutions at 0.5 nmol/L, 1 nmol/L, 5 nmol/L, 10 nmol/L, 20 nmol/L, 50 nmol/L, 100 nmol/L, 200 nmol/L, 500 nmol/L, 1000 nmol/L, 2000 nmol/L. In addition, 100 µL of the internal standard working solution (daidzein, rutin and ($-$)-gallocatechin) of 4000 nmol/L had to be added to each working solution, and the final volume of each working solution was 5 mL. With the concentration ratio of the external standard to the internal standard as the horizontal coordinate and the area ratio of the external standard to the internal standard ratio as the vertical coordinate, standard curves were constructed from the mass spectral peak intensity data of the corresponding quantitative signals of each standard working solution. A good linearity of 204 flavonoid standard curves was determined within the concentration range of 0.5 nmol/L to 200 nmol/L (R2 ≥ 0.9900), and the results are shown in Supplementary Table S3.

2.7. Determination of Qualitative and Quantitative Parameters of the LC-MS/MS Method

A database was constructed based on flavonoid standards, and MS data were analyzed qualitatively. The multiple reaction monitoring (MRM) mode of triple quadrupole MS was used for quantitative analysis, in which precursor ions of the target substance (parent ions) were initially screened by the quadrupole, and ions corresponding to substances with other molecular weights were excluded to limit interference. Precursor ions were then induced to ionization by the collision chamber, broken to form multiple fragment ions and filtered by the triple quadrupole for the selection of fragment ions with the required characteristics, and the interference of non-target ions was simultaneously excluded, resulting in more accurate and reproducible quantification. MS data of honey and nectar samples were obtained, and chromatographic peaks of all targets were integrated and quantitatively analyzed based on the flavonoid standard curves.

2.8. Data Processing

Palynological and physicochemical analysis had six replicates, and flavonoid analysis had three replicates. Mass spectrometry data acquisition was conducted in Analyst 1.6.3 software (AB SCIEX, MA, Framingham, MA, USA). Multiquant 3.0.3 software (AB SCIEX, Framingham, MA, USA) was used for mass spectrometry data procession, and the accuracy of metabolite quantification was referenced to the retention time and peak shape information of the standards and mass spectrometry peaks of the analytes after integral correction. The results are expressed as mean ± standard deviation (SD).

3. Results and Discussion

3.1. Palynological and Physicochemical Characterization of Honey

Firstly, approximately 89.60 ± 2.60% of pollen grains in TCH matched those collected directly from *T. cochinchinensis* plants. Pollen grains from TCH exhibited a prolate shape in the equatorial view (Figure 1C) and a trilobed circular shape in the polar view (Figure 1D), measuring 44 × 22 µm in size with tricolporate and reticulate pattern on the outer wall, thus surpassing the 45% requirement to be considered a monofloral honey [36].

Palynological analysis found that the other 10 types of monofloral honey samples also had a high single pollen rate (Table 1), and the specific pollen morphology is shown in Figure 2.

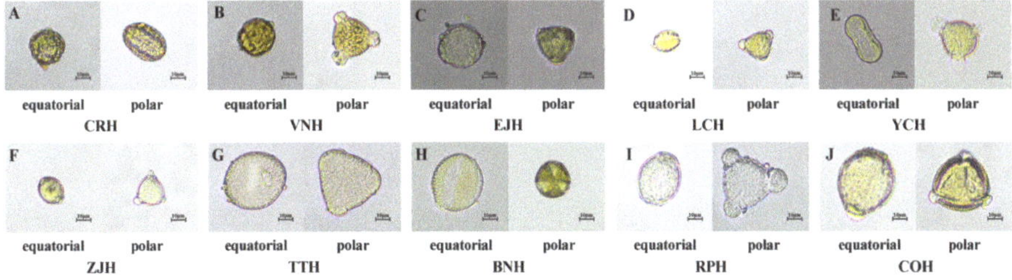

Figure 2. The pollen morphology of 10 common types of Chinese commercial monofloral honey. (**A**): *Citrus reticulata* honey (CRH); (**B**): *Vitex negundo* honey (VNH); (**C**): *Eriobotrya japonica* honey (EJH); (**D**): *Litchi chinensis* honey (LCH); (**E**): *Lycium chinense* honey (YCH); (**F**): *Ziziphus jujuba* honey (ZJH); (**G**): *Tilia tuan* honey (TTH); (**H**): *Brassica napus* honey (BNH); (**I**): *Robinia pseudoacacia* honey (RPH); (**J**): *Camellia oleifera* honey (COH).

Table 1. Native pollen rate of 10 common types of Chinese commercial monofloral honey.

Honey Variety	Native Pollen Rate (%)	Honey Variety	Native Pollen Rate (%)
Citrus reticulata honey (CRH)	75.94 ± 5.22	*Ziziphus jujuba* honey (ZJH)	86.41 ± 2.22
Vitex negundo honey (VNH)	64.81 ± 5.50	*Tilia tuan* honey (TTH)	88.80 ± 1.25
Eriobotrya japonica honey (EJH)	76.82 ± 6.62	*Brassica napus* honey (BNH)	84.85 ± 2.08
Litchi chinensis honey (LCH)	85.51 ± 2.09	*Robinia pseudoacacia* honey (RPH)	71.47 ± 6.49
Lycium chinense honey (YCH)	81.08 ± 2.46	*Camellia oleifera* honey (COH)	87.76 ± 6.63

Table 2 depicts the main physicochemical parameters of TCH. Honey is mostly composed of sugars [37]. The content of total reducing sugars was 74.74% (fructose content, 37.42%; glucose content, 37.32%), and sucrose content was 0.77%, all of which are in accordance with European Union honey standards [38]. The fructose/glucose (F/G) ratio in honey is an important parameter in predicting the crystallization of honey, and when the ratio is <1.11, honey will crystallize very easily [39]. TCH had an (F/G) ratio of 1.00, indicating fast crystallization.

Table 2. Physicochemical parameters of TCH.

Parameter	Mean ± SD	Units
Fructose	37.42 ± 0.71	%
Glucose	37.32 ± 0.36	%
Sucrose	0.77 ± 0.10	%
Water	18.65 ± 0.56	%
HMF	1.87 ± 0.12	mg/kg
Diastase activity	2.55 ±0.32	mL/(g·h)
Electrical conductivity	0.14 ± 0.003	mS/cm
ash content	0.07 ±0.001	g/100 g
Color value	32.00 ± 0.00	mm Pfund
Free acidity	11.82 ± 0.22	mL/kg
Pollen grains concentration	18,025.00 ± 641.67	grain/mL
Fe	6.23 ± 0.05	mg/kg
Cu	105.31 ± 5.98	µg/kg
Zn	5.41 ± 0.29	mg/kg

Water content is one of the significant parameters used to identify the quality of honey and is taken as a vital indicator for maturity, viscosity and stability [40]. The water content of TCH was 18.65%, which matched the standards that the water content should not exceed 20% [38].

Moreover, 5-HMF is a cyclic aldehyde, an intermediate product from the Maillard reaction (a non-enzymatic browning reaction) during processing or long storage of honey. 5-HMF content is widely recognized as a parameter affecting honey freshness [41]. The content of 5-HMF was 1.87 mg/kg, which is within the acceptable range for honey samples based on European Union [38].

Diastases play important roles in the process of honey maturation, whose function is to digest the starch molecule in a mixture of maltose and maltotriose [42,43]. Diastase content depends on the different floral and geographical origins of the honey. TCH exhibited a diastase activity of 2.55 mL/ (g h), which is consistent with the findings of Liu et al. who described naturally low diastase activity in TCH [22].

Trace minerals are important constituents of honey and play specific roles in human health [44]. The presence of Fe in honey can alleviate anaemia and increase immunity in honey eaters [45]. Cu is necessary for normal human health and growth and contributes to immune function [46]. Zn is an essential antioxidant mineral that can promote wound healing and decrease risks of cancer and cardiovascular diseases [47]. The mineral concentrations of Fe, Cu and Zn in TCH were 6.23 mg/kg, 105.31 µg/kg and 5.41 mg/kg, respectively. According to a previous study conducted by our research group, TCH exhibited high contents of Fe and Zn compared to other nine types of honey [48].

Taken together, these findings indicate that TCH can be considered a high-quality honey with a unique flavor.

3.2. Screening and Identification of Unique Flavonoid Markers in TCH

The class of flavonoids comprises over 50% of phenolic compounds, which are the essential products of secondary plant metabolism [43]. The flavonoid composition of honey is mainly associated with its floral source and geographical origin [49], thus serving as a tool for honey classification and authentication, particularly for monofloral honeys [50].

A total of 35 flavonoids were detected in TCH using LC-MS/MS (Figure 3A). Compared to the other 11 types of honey tested herein, TCH was found to contain 12 distinctive flavonoids, which was also the largest number among all tested honeys. These 12 flavonoids included 5,7,3′,4′-tetramethoxyflavone, genistin, neohesperidin, mangiferin, epigallocatechin, oroxinA, catechingallate, sieboldin, (−)-gallocatechin gallate, dihydromyricetin, myricitrin and naringenin-7-glucoside (Figure 3B).

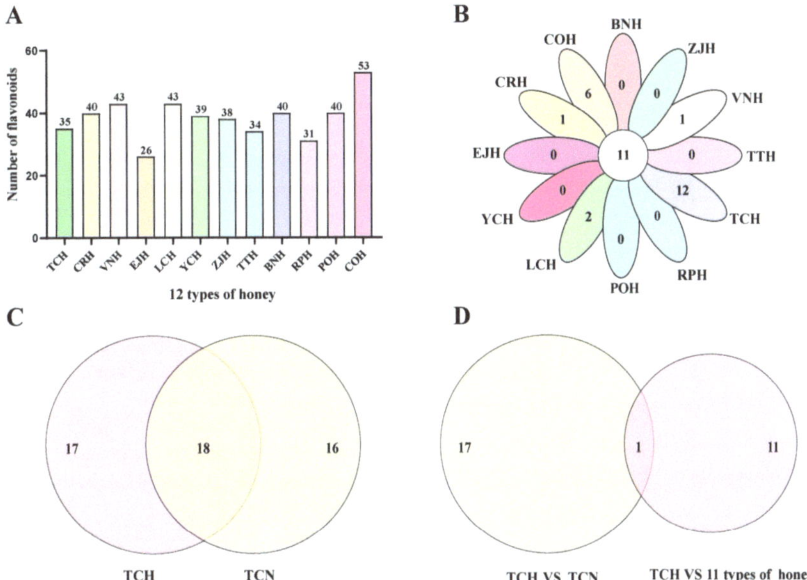

Figure 3. (**A**) Flavonoid species identified in TCH and 11 other common types of Chinese commercial honey. (**B**) Flavonoid species distinctive in one type of honey relative to 11 other common types of Chinese commercial honey. TCH has 12 distinctive flavornoid species. (**C**) The flavonoid species that are shared in TCH and TCN. (**D**) Flavonoids distinctive to TCH and TCN relative to 11 other common types of Chinese commercial honey.

In addition, 18 flavonoids were shared between TCH and TCN (Figure 3C). A comparative analysis between the 18 flavonoids and the 12 flavonoids previously found to be unique to TCH among the 12 types of honey revealed that only one flavonoid, (−)-gallocatechin gallate (GCG), was uniquely found in both TCH and TCN (Figure 3D). Therefore, GCG was considered a distinct flavonoid marker of TCH. Chromatographic and MS spectra of GCG in TCH and TCN, as well as of the GCG standard, are shown in Figure 4.

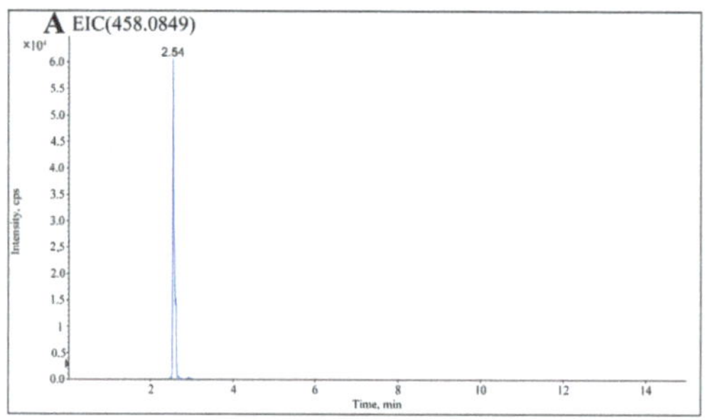

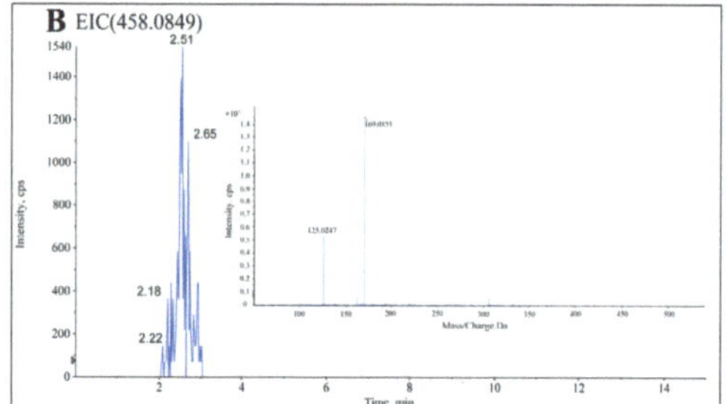

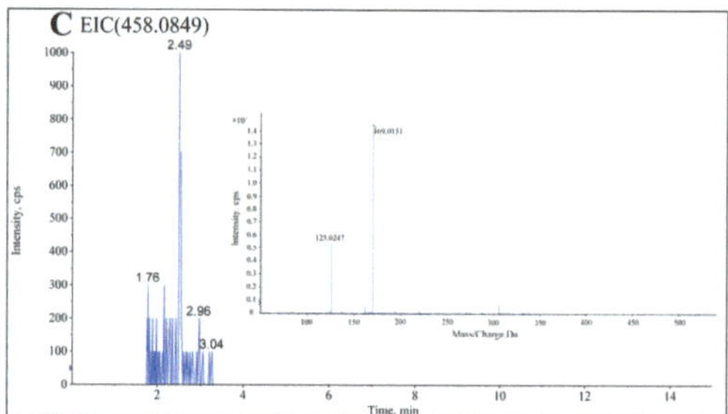

Figure 4. The typical chromatograms of the extracts from the positive/negative mode analyzed by LC-MS/MS. (**A**) Extracted ion chromatogram (EIC) of GCG standard. (**B**) EIC and mass spectrum of GCG in TCN. (**C**) EIC and mass spectrum of GCG in TCH.

3.3. Method Validation and Quantification of (−)-Gallocatechin Gallate

GCG, a catechin compound, has various health benefits, including antioxidant and antibacterial properties [51], cholesterol- and triglyceride-lowering effects [52], melanin synthesis inhibition [53] and neuroprotective and cardioprotective effects [54,55]. Thus, detecting the accurate quantitative measurement of GCG in TCH would allow evaluating its authenticity and also contribute to popularizing this characteristic monofloral honey.

Herein, an LC-MS/MS method was developed to quantify GCG in TCH and TCN. To achieve this, 11 GCG standard solutions were used to construct a standard curve, allowing the detection of GCG content in TCH and TCN samples. The GCG standard curve was described by the equation $y = 1010.75160x − 2834.40065$ ($R^2 > 0.99$) within the linear range of 5–2000 nmol/L. The limits of detection (LOD), estimated to a signal to noise (S/N) ratio of 3, and the limits of quantification (LOQ), estimated to a signal to noise (S/N) ratio of 10, for GCG were 1.14 nmol/kg and 3.43 nmol/kg, respectively. Based on the standard curve, the content of GCG in TCH and TCH was 130.78 ± 4.44 nmol/kg and 96.33 ± 2.16 nmol/kg, respectively. Moreover, the relative standard deviation (RSD) of TCH and TCN was 3.40% and 2.25%, respectively. The combined results are shown in Table 3. Taken together, the LC-MS/MS method developed herein could be considered sensitive and reliable for the detection of GCG in TCH and TCN.

Table 3. GCG of standard curve, LOD, LOQ and the content of GCG in TCH and TCN.

Compound	Standard Curve	LOD (nmol/kg)	LOQ (nmol/kg)	Regression (R^2)	TCH ($n = 3$)		TCN ($n = 3$)	
					Content (nmol/kg)	RSD (%)	Content (nmol/kg)	RSD (%)
GCG	$y = 1010.75160x − 2834.40065$	1.14	3.43	0.9994	130.78 ± 4.44	3.40	96.33 ± 2.16	2.25

4. Conclusions

In this study, the LC-MS/MS method was applied to identify flavonoids in TCH, TCN and 11 other types of commonly commercial Chinese honey to accurately identify and quantify the characteristic markers of TCH. GCG was identified as a unique flavonoid marker for TCH. In addition, a reliable and accurate LC-MS/MS method was established for the first time to identify GCG in TCH and TCN. Thus, the findings of the present study provide a novel and reliable solution for the authentication and quality control of TCH, which would also provide theoretical support for developing standards for TCH products.

Supplementary Materials: The following supporting information can be downloaded at: https://www.mdpi.com/article/10.3390/foods13121879/s1, Table S1: Standard information for 204 flavonoids; Table S2: The specific flavonoid standards monitored in positive and negative ion modes; Table S3: Standard working curve information for 204 flavonoids.

Author Contributions: Conceptualization, Z.Z.; methodology, H.J. and Z.L.; validation, H.J. and Z.L.; formal analysis, S.Z.; data curation S.Z.; writing—original draft preparation, H.J. and Z.L.; writing—review and editing, H.J. and Z.L.; visualization, S.Z.; supervision, Z.Z.; project administration, Z.Z.; funding acquisition, Z.Z. All authors have read and agreed to the published version of the manuscript.

Funding: This research was funded by the National Key R&D Program Project of China (Grant Number: 2022YFD1600202) and the Earmarked Fund for the China Agricultural Research System (Grant Number: CARS-44-KXJ15).

Data Availability Statement: The original contributions presented in the study are included in the article/supplementary material, further inquiries can be directed to the corresponding author.

Conflicts of Interest: The authors declare no conflicts of interest.

References

1. Becerril-Sánchez, A.L.; Quintero-Salazar, B.; Dublán-García, O.; Escalona-Buendía, H.B. Phenolic Compounds in Honey and Their Relationship with Antioxidant Activity, Botanical Origin, and Color. *Antioxidants* **2021**, *10*, 1700. [CrossRef] [PubMed]
2. Bora, F.D.; Andrecan, A.F.; Călugăr, A.; Bunea, C.I.; Popescu, M.; Petrescu-Mag, I.V.; Bunea, A. Comprehensive Elemental Profiling of Romanian Honey: Exploring Regional Variance, Honey Types, and Analyzed Metals for Sustainable Apicultural and Environmental Practices. *Foods* **2024**, *13*, 1253. [CrossRef] [PubMed]
3. Massaro, A.; Zacometti, C.; Bragolusi, M.; Buček, J.; Piro, R.; Tata, A. Authentication of the botanical origin of monofloral honey by dielectric barrier discharge ionization high resolution mass spectrometry (DBDI-HRMS). Breaching the 6 s barrier of analysis time. *Food Control* **2024**, *160*, 110330. [CrossRef]
4. Wang, H.; Li, L.; Lin, X.; Bai, W.; Xiao, G.; Liu, G. Composition, functional properties and safety of honey: A review. *J. Sci. Food Agric.* **2023**, *103*, 6767–6779. [CrossRef] [PubMed]
5. Ciucure, C.T.; Geană, E.I. Phenolic compounds profile and biochemical properties of honeys in relationship to the honey floral sources. *Phytochem. Anal.* **2019**, *30*, 481–492. [CrossRef] [PubMed]
6. Lawag, I.L.; Lim, L.-Y.; Joshi, R.; Hammer, K.A.; Locher, C. A comprehensive survey of phenolic constituents reported in Monofloral honeys around the globe. *Foods* **2022**, *11*, 1152. [CrossRef] [PubMed]
7. Palma-Morales, M.; Huertas, J.R.; Rodríguez-Pérez, C. A Comprehensive Review of the Effect of Honey on Human Health. *Nutrients* **2023**, *15*, 3056. [CrossRef]
8. Machado, A.M.; Miguel, M.G.; Vilas-Boas, M.; Figueiredo, A.C. Honey Volatiles as a Fingerprint for Botanical Origin—A Review on their Occurrence on Monofloral Honeys. *Molecules* **2020**, *25*, 374. [CrossRef] [PubMed]
9. Schanzmann, H.; Augustini, A.; Sanders, D.; Dahlheimer, M.; Wigger, M.; Zech, P.M.; Sielemann, S. Differentiation of Monofloral Honey Using Volatile Organic Compounds by HS-GCxIMS. *Molecules* **2022**, *27*, 7554. [CrossRef]
10. Alvarez-Suarez, J.M.; Gasparrini, M.; Forbes-Hernández, T.Y.; Mazzoni, L.; Giampieri, F. The Composition and Biological Activity of Honey: A Focus on Manuka Honey. *Foods* **2014**, *3*, 420–432. [CrossRef]
11. Lin, B.; Daniels, B.J.; Middleditch, M.J.; Furkert, D.P.; Brimble, M.A.; Bong, J.; Stephens, J.M.; Loomes, K.M. Utility of the *Leptospermum scoparium* Compound Lepteridine as a Chemical Marker for Manuka Honey Authenticity. *ACS Omega* **2020**, *5*, 8858–8866. [CrossRef] [PubMed]
12. Ren, C.; Wang, K.; Luo, T.; Xue, X.; Wang, M.; Wu, L.; Zhao, L. Kaempferol-3-O-galactoside as a marker for authenticating *Lespedeza bicolor* Turcz. monofloral honey. *Food Res. Int.* **2022**, *160*, 111667. [CrossRef] [PubMed]
13. Li, Z.; Huang, Q.; Zheng, Y.; Zhang, Y.; Liu, B.; Shi, W.K.; Zeng, Z.J. Kaempferitrin: A Flavonoid Marker to Distinguish *Camellia oleifera* Honey. *Nutrients* **2023**, *15*, 435. [CrossRef]

14. Zhao, L.W.; Ren, C.J.; Xue, X.F.; Lu, H.X.; Wang, K.; Wu, L.M. Safflomin A: A novel chemical marker for *Carthamus tinctorius* L. (Safflower) monofloral honey. *Food Chem.* **2022**, *366*, 130584. [CrossRef]
15. Zhao, T.; Zhao, L.; Wang, M.; Qi, S.; Xue, X.; Wu, L.; Li, Q. Identification of characteristic markers for monofloral honey of *Astragalus membranaceus* var. *mongholicus* Hsiao: A combined untargeted and targeted MS-based study. *Food Chem.* **2023**, *404*, 134312.
16. Esser, H.-J. A revision of *Triadica* Lour. (euphorbiaceae). *Harv. Pap. Bot.* **2002**, *7*, 17–21.
17. Wang, S.Q.; Chen, Y.Y.; Yang, Y.C.; Wu, W.; Liu, Y.; Fan, Q.; Zhou, R.C. Phylogenetic relationships and natural hybridization in *Triadica* inferred from nuclear and chloroplast DNA analyses. *Biochem. Syst. Ecol.* **2016**, *64*, 142–148. [CrossRef]
18. Huang, Z.H. *DNA Barcode Sequence of Lingnan Chinese Herbs*; China Medical Science and Technology Press: Beijing, China, 2017; p. 219.
19. Sabandar, C.; Jalil, J.; Ahmat, N.; Aladdin, N.-A. Assessment of antioxidant and xanthine oxidase inhibitory activity of *Triadica cochinchinensis* stem bark. *Curr. Res. Biosci. Biotechnol.* **2019**, *1*, 39–44. [CrossRef]
20. He, Q.B.; Zhang, L.; Li, T.; Li, C.H.; Song, H.N.; Fan, P.H. Genus *Sapium* (Euphorbiaceae): A review on traditional uses, phytochemistry, and pharmacology. *J. Ethnopharmacol.* **2021**, *277*, 114206. [CrossRef]
21. Luo, L.; Zhang, J.; Liu, M.; Qiu, S.; Yi, S.; Yu, W.; Liu, T.; Huang, X.; Ning, F. Monofloral *Triadica cochinchinensis* Honey Polyphenols Improve Alcohol-Induced Liver Disease by Regulating the Gut Microbiota of Mice. *Front. Immunol.* **2021**, *12*, 673903. [CrossRef]
22. Liu, T.; Qiao, N.; Ning, F.J.; Huang, X.Y.; Luo, L.P. Identification and characterization of plant-derived biomarkers and physicochemical variations in the maturation process of *Triadica cochinchinensis* honey based on UPLC-QTOF-MS metabolomics analysis. *Food Chem.* **2023**, *408*, 135197. [CrossRef] [PubMed]
23. Zhang, J.P. The Mechanism of *Triadica cochinchinensis* Honey Ameliorating Alcohol Liver Damage of Mice by Adjusting the Gut Microbiota. Master's Thesis, Nanchang University, Nanchang, China, 2023.
24. Luo, L.P.; Qiao, N.; Guo, L.M.; Liu, T. Analysis of volatile components of *Triadica cochinchinensis* honey by HS-SPME-GC-MS. *J. Nanchang Univ.* **2022**, *46*, 320–326+333.
25. Zeng, Z.J. *Apiculture*, 4th ed.; China Agricultural Press: Beijing, China, 2023; p. 23.
26. Li, Q.; Wu, L. Bee Products: The Challenges in Quality Control. *Foods* **2023**, *12*, 3699. [CrossRef]
27. Wang, X.; Chen, Y.; Hu, Y.; Zhou, J.; Chen, L.; Lu, X. Systematic review of the characteristic markers in honey of various botanical, geographic, and entomological origins. *ACS Food Sci. Technol.* **2022**, *2*, 206–220. [CrossRef]
28. Qi, D.; Lu, M.; Li, J.; Ma, C. Metabolomics Reveals Distinctive Metabolic Profiles and Marker Compounds of Camellia (*Camellia sinensis* L.) Bee Pollen. *Foods* **2023**, *12*, 2661. [CrossRef]
29. Ranneh, Y.; Akim, A.M.; Hamid, H.A.; Khazaai, H.; Fadel, A.; Zakaria, Z.A.; Albujja, M.; Bakar, M.F.A. Honey and its nutritional and anti-inflammatory value. *BMC Complement. Med. Ther.* **2021**, *21*, 30. [CrossRef] [PubMed]
30. Truchado, P.; Ferreres, F.; Tomas-Barberan, F.A. Liquid chromatography–tandem mass spectrometry reveals the widespread occurrence of flavonoid glycosides in honey, and their potential as floral origin markers. *J. Chromatogr. A* **2009**, *1216*, 7241–7248. [CrossRef] [PubMed]
31. Song, X.Y.; Yao, Y.F.; Yang, W.D. Pollen analysis of natural honeys from the central region of Shanxi, North China. *PLoS ONE* **2012**, *7*, e49545. [CrossRef] [PubMed]
32. Yang, S.; Mao, L.; Zheng, Z.; Chen, B.; Li, J. Pollen atlas for selected subfamilies of Euphorbiaceae from Southern China: A complementary contribution to Quaternary pollen analysis. *Palynology* **2020**, *44*, 659–673. [CrossRef]
33. Xu, W. *Nectar and Pollen Plants of China*; Heilongjiang Science & Technology Press: Harbin, China, 1992.
34. AOAC. *International Official Methods of Analysis of AOAC International*, 20th ed.; AOAC: Gaithersburg, MD, USA, 2016.
35. Zhou, J.; Suo, Z.; Zhao, P.; Cheng, N.; Gao, H.; Zhao, J.; Cao, W. Jujube honey from China: Physicochemical characteristics and mineral contents. *J. Food Sci.* **2013**, *78*, C387–C394. [CrossRef]
36. Bicudo de Almeida-Muradian, L.; Monika Barth, O.; Dietemann, V.; Eyer, M.; Freitas, A.d.S.d.; Martel, A.-C.; Marcazzan, G.L.; Marchese, C.M.; Mucignat-Caretta, C.; Pascual-Maté, A. Standard methods for *Apis mellifera* honey research. *J. Apic. Res.* **2020**, *59*, 1–62. [CrossRef]
37. Živkov Baloš, M.; Popov, N.; Jakšić, S.; Mihaljev, Ž.; Pelić, M.; Ratajac, R.; Ljubojević Pelić, D. Sunflower Honey—Evaluation of Quality and Stability during Storage. *Foods* **2023**, *12*, 2585. [CrossRef] [PubMed]
38. Council, E. Council Directive 2001/110/EC of 20 December 2001 relating to honey. *Off. J. Eur. Communities L* **2002**, *10*, 47–52.
39. Wu, L.; Du, B.; Vander Heyden, Y.; Chen, L.; Zhao, L.; Wang, M.; Xue, X. Recent advancements in detecting sugar-based adulterants in honey—A challenge. *TrAC Trends Anal. Chem.* **2017**, *86*, 25–38. [CrossRef]
40. Serin, S.; Turhan, K.N.; Turhan, M. Correlation between water activity and moisture content of Turkish flower and pine honeys. *Ciência Tecnol. Aliment.* **2018**, *38*, 238–243. [CrossRef]
41. Shapla, U.M.; Solayman, M.; Alam, N.; Khalil, M.I.; Gan, S.H. 5-Hydroxymethylfurfural (HMF) levels in honey and other food products: Effects on bees and human health. *Chem. Cent. J.* **2018**, *12*, 35. [CrossRef]
42. De-Melo, A.A.M.; Almeida-Muradian, L.B.D.; Sancho, M.T.; Pascual-Maté, A. Composition and properties of *Apis mellifera* honey: A review. *J. Apic. Res.* **2018**, *57*, 5–37. [CrossRef]
43. Da Silva, P.M.; Gauche, C.; Gonzaga, L.V.; Costa, A.C.; Fett, R. Honey: Chemical composition, stability and authenticity. *Food Chem.* **2016**, *196*, 309–323. [CrossRef] [PubMed]

44. Solayman, M.; Islam, M.A.; Paul, S.; Ali, Y.; Khalil, M.I.; Alam, N.; Gan, S.H. Physicochemical properties, minerals, trace elements, and heavy metals in honey of different origins: A comprehensive review. *Compr. Rev. Food Sci. Food Saf.* **2016**, *15*, 219–233. [CrossRef]

45. Lestari, I.; Lestari, A.E. *Effectiveness of Honey in Increasing Hemoglobin Levels of Mothers Post Sectio Caesarea*; Humanistic Network for Science and Technology: Washington, DC, USA, 2019.

46. Al-Waili, N.S. Effects of daily consumption of honey solution on hematological indices and blood levels of minerals and enzymes in normal individuals. *J. Med. Food* **2003**, *6*, 135–140. [CrossRef]

47. Chasapis, C.T.; Ntoupa, P.-S.A.; Spiliopoulou, C.A.; Stefanidou, M.E. Recent aspects of the effects of zinc on human health. *Arch. Toxicol.* **2020**, *94*, 1443–1460. [CrossRef] [PubMed]

48. Jiang, H.Z.; Jiang, W.J.; Li, Z.; Zhong, S.Q.; He, X.J.; Yan, W.Y.; Xi, F.G.; Wu, W.M.; Zeng, Z.J. Production and composition analysis of high quality *Triadica cochinchinensis* honey. *Acta Agric. Univ. Jiangxiensis* **2023**, *45*, 1473–1485.

49. Masad, R.J.; Haneefa, S.M.; Mohamed, Y.A.; Al-Sbiei, A.; Bashir, G.; Fernandez-Cabezudo, M.J.; Al-Ramadi, B.K. The Immunomodulatory Effects of Honey and Associated Flavonoids in Cancer. *Nutrients* **2021**, *13*, 1269. [CrossRef] [PubMed]

50. Cianciosi, D.; Forbes-Hernández, T.Y.; Afrin, S.; Gasparrini, M.; Reboredo-Rodriguez, P.; Manna, P.P.; Zhang, J.; Bravo Lamas, L.; Martínez Flórez, S.; Agudo Toyos, P.; et al. Phenolic Compounds in Honey and Their Associated Health Benefits: A Review. *Molecules* **2018**, *23*, 2322. [CrossRef] [PubMed]

51. Li, K.K.; Zhou, X.L.; Liu, C.L.; Yang, X.R.; Han, X.Q.; Shi, X.G.; Song, X.H.; Ye, C.X.; Ko, C.H. Preparative separation of gallocatechin gallate from *Camellia ptilophylla* using macroporous resins followed by sephadex LH-20 column chromatography. *J. Chromatogr. B* **2016**, *1011*, 6–13. [CrossRef] [PubMed]

52. Lee, S.M.; Kim, C.W.; Kim, J.K.; Shin, H.J.; Baik, J.H. GCG-rich tea catechins are effective in lowering cholesterol and triglyceride concentrations in hyperlipidemic rats. *Lipids* **2008**, *43*, 419–429. [CrossRef] [PubMed]

53. Wang, W.; Di, T.; Wang, W.; Jiang, H. EGCG, GCG, TFDG, or TSA Inhibiting Melanin Synthesis by Downregulating MC1R Expression. *Int. J. Mol. Sci.* **2023**, *24*, 11017. [CrossRef]

54. Hirai, M.; Hotta, Y.; Ishikawa, N.; Wakida, Y.; Fukuzawa, Y.; Isobe, F.; Nakano, A.; Chiba, T.; Kawamura, N. Protective effects of EGCG or GCG, a green tea catechin epimer, against postischemic myocardial dysfunction in guinea-pig hearts. *Life Sci.* **2007**, *80*, 1020–1032. [CrossRef]

55. Park, D.H.; Park, J.Y.; Kang, K.S.; Hwang, G.S. Neuroprotective effect of gallocatechin gallate on glutamate-induced oxidative stress in hippocampal ht22 cells. *Molecules* **2021**, *26*, 1387. [CrossRef]

Article

Assessing Mineral Content and Heavy Metal Exposure in Abruzzo Honey and Bee Pollen from Different Anthropic Areas

Federica Flamminii [1,*], Ada Consalvo [2], Angelo Cichelli [1] and Alessandro Chiaudani [1]

1 Department of Innovative Technologies in Medicine and Dentistry, University "G. d'Annunzio" of Chieti-Pescara, Via dei Vestini, 66100 Chieti, Italy; angelo.cichelli@unich.it (A.C.); achiaudani@unich.it (A.C.)
2 Center for Advanced Studies and Technology (CAST), "G. d'Annunzio" University of Chieti-Pescara, Via Luigi Polacchi, 11, 66100 Chieti, Italy; ada.consalvo@libero.it
* Correspondence: federica.flamminii@unich.it

Abstract: Honey and bee pollen offer potential health benefits due to their nutrient and bioactive molecules, but they may also harbor contaminants such as heavy metals. This study aimed to assess the content of different metals, including Mg, Al, K, Ca, V, Cr, Mn, Fe, Co, Ni, Zn, Cu, As, Rb, Sr, Cd, Cs, Tl, Pb and U, in honey and bee pollen collected from different Abruzzo region (Italy) areas (A1, A2, A3, A4), characterized by different anthropic influences described by Corine Land Cover maps. Differences were observed in the mineral and heavy metal content associated with the influence of biotic and abiotic factors. Honeys were found to be safe in regard to non-carcinogenic risk in all the consumer categories ($THQ_m < 1$). A particular carcinogenic risk concern was identified for toddlers associated with Cr ($LCTR > 1 \times 10^{-4}$) in A1, A2 and A3 apiaries. Pb and Ni represent potential non-carcinogenic and carcinogenic health risks in children and adults due to bee pollen consumption, showing high values of THQ_m and LCTR. The results suggest the advantages of utilizing bee products to screen mineral and heavy metal content, providing valuable insights into environmental quality and potential health risks.

Keywords: honey; bee pollen; ICP-MS; heavy metal; risk assessment; environmental impact

Citation: Flamminii, F.; Consalvo, A.; Cichelli, A.; Chiaudani, A. Assessing Mineral Content and Heavy Metal Exposure in Abruzzo Honey and Bee Pollen from Different Anthropic Areas. *Foods* **2024**, *13*, 1930. https://doi.org/10.3390/foods13121930

Academic Editors: Liming Wu and Qiangqiang Li

Received: 8 May 2024
Revised: 11 June 2024
Accepted: 14 June 2024
Published: 19 June 2024

1. Introduction

Since ancient times, throughout history, honey has accompanied human beings, serving as a vital food source and a key element in religious, mystical, and medicinal practices [1].

Honey, produced by *Apis mellifera* bees, comes from the nectar of flowers, plant secretions, and aphids' honeydew. Bees collect, transform, dehydrate, store, and mature it in the honeycomb [2]. It is rich in nutrients, predominantly carbohydrates, which make up about 75% of its composition, with glucose and fructose being the main sugars (85–95%). Additionally, honey contains traces of organic acids, enzymes, amino acids, and pigments. The water content ranges from 10 to 25%, while minerals, varying by the honey's origin, constitute 0.04%–0.2%. Key minerals include potassium (K) (up to 70%), calcium (Ca) and sodium (Na), magnesium (Mg), iron (Fe), manganese (Mn), and zinc (Zn), with the average contents exceeding 1 mg kg^{-1} [3–6].

Along with its nutritional properties, honey brings several functional and health benefits (antimicrobial, antioxidant, anti-inflammatory, antidiabetic, wound healing, anticancer, anti-proliferative, immunomodulatory effects, gastrointestinal tract diseases, cardiovascular effects, ophthalmology) well documented and described by Aga and coauthors [7].

The worldwide production of honey in 2021 reached 1772 M ton^{-1}, and about 22% was produced in Europe. In Italy, 23.000 t of honey were produced in 2022, and about 3% of this amount (690 t) was produced in the Abruzzo region [8].

Bee pollen is harvested by honeybees from plant flowers and enriched with salivary enzymes and nectar to obtain small granular-looking grains (bee pollen) that are transported into the apiary [9]. It contains carbohydrates (13–55%), proteins (10–40%), lipids (1–13%)

and fibers (0.3–20%), with a moisture content varying from 4 to 8%. Additionally, bee pollen has a high mineral content (2.5–6.5 g/100 g), predominantly potassium (K) (about 60% of total mineral content), along with magnesium (Mg), sodium (Na) and calcium (Ca), ranging from 10% to 20% [10–15]. Bee pollen is also rich in secondary metabolites, including biotin, folic acid, carotenoid pigments, niacin, phytosterols, polyphenols, thiamine, tocopherol, flavonoids, sterols, terpenes, vitamins, enzymes, and coenzymes. Consuming daily doses of 20–40 g of bee pollen can provide recommended daily intakes (RDIs) for various elements at notably high levels.

Bee pollen is a natural strength supplement to the body's immune and physiological systems, making it attractive for use in the diets of children and adults suffering from certain avitaminoses and loss of appetite. It improves blood supply to the nerve tissue, powers mental performance, and reduces the state of fatigue while having a positive effect on the liver, heart, prostate, and allergy diseases. The primary consumers of bee pollen include advocates of health-conscious and environmentally friendly lifestyles, as well as the elderly, due to its antioxidant and other therapeutic effects [10–13].

Mineral elements contained in honey and bee pollen could be both essential and non-essential to human biological functions. An inadequate dietary deficiency of the essential mineral elements results in a variety of diseases or syndromes; conversely, these mineral elements can become harmful in excessive amounts [14,15]. Non-essential minerals can exert toxicity, even at low concentrations, and can affect the level of the essential elements in the body [16,17]. The essential elements include macrominerals (sodium, magnesium, phosphorous, sulfur, chlorine, potassium, and calcium) and trace elements (silicon, vanadium, chromium, manganese, iron, cobalt, nickel, copper, zinc, selenium, molybdenum and iodine).

The macrominerals, with an average content exceeding 1 mg kg^{-1} [18], are responsible for the maintenance of the ionic balance of structural skeletal compounds, amino acids, and nucleic acids. Trace elements have several physiological and biochemical functions for the correct cellular metabolism, influencing the circulatory system and reproduction and composing structural proteins, hormones, and key enzymes, e.g., zincin, iron in hemoglobin, and selenium in glutathione peroxidase enzyme [19–22].

Certain non-essential elements, such as aluminum, vanadium, arsenic, rubidium, strontium, cadmium, cesium, thallium, lead, and uranium, may contaminate honey and pollen, and among the most potentially toxic are heavy metals characterized by a high atomic weight (over 63.5 and with a specific gravity higher than 5.0). The major elements included in this class are as follows: Pb, Cd, Co, Cr, Cu, Fe, As, Ni, Zn, and Hg [23]. In general, they are found naturally on the Earth's crust, but their growing utilization results in an increase in metallic substances in both the terrestrial and aquatic environments [24]. The primary pollution sources are the metal-based industries, leaching of metals from landfills, waste dumps, excretion, livestock and chicken manure, runoffs, automobiles, and roadworks. The use of pesticides, insecticides, and fertilizers in agriculture are the secondary source of heavy metal pollution [25]. Also, natural causes, such as volcanic activity, metal corrosion, metal evaporation from soil and water and sediment re-suspension, soil erosion, and geological weathering can increase heavy metal pollution.

Heavy metals interact with nuclear proteins together with DNA, causing site-specific direct and indirect damages. In the first case, conformational changes occur to the biomolecules, while the second is a result of the production of reactive oxygen and nitrogen species such as hydroxyl and superoxide radicals, hydrogen peroxide, nitric oxide, and other endogenous oxidants. These toxic elements can lead to acute damage to many vital organs, like the kidneys, liver, brain, etc. In addition, prolonged exposure can trigger blood pressure alteration, anxiety, and passivity disorders. According to the US Environmental Protection Agency (USEPA) and the International Agency for Research on Cancer (IARC), arsenic, cadmium, chromium, lead, and mercury are the most dangerous metals and are also classified as either "known" or "probable" human carcinogens [21]. Chromium and nickel can exert toxicity, affecting the respiratory system and inducing carcinogenesis, allergies, infection diseases, and intestinal microbes [26,27]. The consumption of arsenic element in food products leads to skin lesions and cardiovascular disturbances, while cadmium is also responsible for placental damage, prostate cancer, and renal lesions; moreover, ingestion of mercury can cause cerebral palsy and mental retardation. Children are vulnerable to lead, with it particularly affecting their brain and peripheral nervous system development, while hypertension and kidney damage were observed in adults upon its prolonged consumption [22,28,29]. The latter is the only one to date that has a maximum limit in honey set by law, set at 0.10 mg kg^{-1} wet weight, as established by the Commission Regulation [30].

In general, the content of individual elements can vary considerably among the different honey and pollen taxa. This variation depends on the plant species that bees forage as well as on the landscape and the different morphological characteristics that influence the chemical composition of the ground that surrounds the beehives, which, in turn, is influenced by the levels of environmental pollution [31–33]. The need to monitor and protect the environment more carefully and detect the sources of pollution have become highly topical issues. In recent years, environmental monitoring with bees has assumed greater importance due to its characteristics of management simplicity and cost-effectiveness. Due to their extensive flight range of approximately 1.5 to over 3 km from their hive, covering an estimated area of about 7–28 km^2 (700–2800 ha) [13,34], bees serve as dynamic environmental sensors, unlike many other largely immobile bioindicators [35–37]. The mortality rates of these bioindicators correlate with the levels of environmental pollution, making them valuable for detecting traces of harmful pollutants such as agricultural pesticides, antibiotics from human and livestock sources, heavy metals, radionuclides, pathogenic microorganisms, and other contaminants [38]. Moreover, bees function as "biofilters", mitigating contamination levels in nectar during honey production, even in environments with high pollution levels [39–42].

As a result, honey and pollen, being susceptible to environmental influences, can be significant sources of chemical contaminant exposure, posing potential public health concerns. However, because bee pollen undergoes less of a transformation by bees, it tends to provide a more accurate reflection of environmental contamination [43]. In this regard, some authors have suggested that pollen may serve as a superior bioindicator of environmental pollution compared to honey, which is most effective as a bioindicator in cases of significant contamination [44].

The aim of the present study was to examine the mineral composition of polyfloral honey collected from various areas within the Abruzzo region. Additionally, the investigation was extended to bee pollen obtained from the same hives, aiming to evaluate the impact of the "natural" filtration process carried out by the bees. The selected sampling areas were categorized based on their urban, agricultural, and natural environments, representing a territory model. Furthermore, based on the elemental composition data, the study conducted a risk assessment to evaluate both the carcinogenic and non-carcinogenic risks associated with the ingestion of honey for toddlers, children, adolescents, and adults as well as the ingestion of pollen for children and adults, highlighting the diverse implications associated with their consumption.

2. Materials and Methods

2.1. Sample Collection

Eight honey samples and nine bee pollen samples were collected in the following four suburban areas of the Abruzzo region: the Chieti province (Ortona, Filetto and Frisa municipalities, Italy) and the Teramo province (the Capsano district in the Penna S. Andrea municipality, Italy). Apiaries were coded as follows: A1: Ortona, A2: Filetto, A3, Frisa, A4, S. Andrea. Figure 1 shows the geographical locations of the sampling sites (Figure 1a) and the flight area of the beehives (Figure 1b).

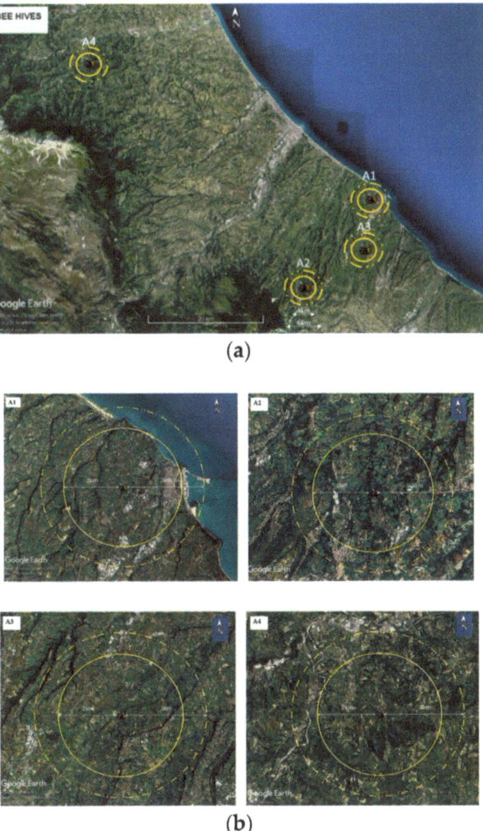

Figure 1. (**a**) Geographical location of the beehives in Abruzzo region and (**b**) the flight area of each beehive (**A1–A4**). The images are adapted from those obtained from Google Earth Image © Airbus 2024 Image © TerraMetrics 2024.

Honey samples were gathered during the periods of May–June 2019 and June–July 2020, whereas bee pollen samples were collected in July 2020 and September 2020. Fresh polyfloral honey and bee pollen were directly procured from beekeepers and subsequently stored in laboratory-grade bottles at a temperature of $-20\,^{\circ}\mathrm{C}$ until analysis.

2.2. Corine Land Cover Use of Soil

To deepen the spatial pressure of land use on forage areas, Google Earth and Corine Land Cover 2018 (CLC) project map information tools were used. The Corine Vector soil data viewer codes appropriately different land-use types, with a 25 ha/100 m minimum mapping unit (Copernicus Land Monitoring Service [45]. The description of the codes is reported in the Supplementary Materials (Table S1). The soil use maps were produced around the beehives for a flight area of 12.5 km^2 (2 km from their hive).

2.3. Sample Preparation

Honey and pollen samples were digested using a previously described procedure [46,47]. Acidic digestion of 0.1 g of pollen samples was performed in sterile polystyrene tubes (15 mL, BD FalconTM, BD Biosciences, Franklin Lakes, NJ, USA) by adding 0.750 mL 69% (v/v) HNO$_3$, heating at 60 °C O/N, and subsequently by adding 0.500 mL of 30% (v/v) H$_2$O$_2$ with final heating at 60 °C for 8 h. Acidic digestion of 1 g of honey samples was performed by adding of 1 g of 18.2 MΩ cm^{-1} deionized water and 2 mL of 69% (v/v) HNO$_3$ heating at 60 °C for 8 h. The digested pollen and honey samples were diluted to a final volume of 12 mL and 8 mL, respectively, with 18.2 MΩ cm^{-1} deionized water, and they were analyzed by Inductively Coupled Plasma Mass Spectrometry (ICP-MS). The external standard method was applied for quantification, and we later verified the accuracy of the method with fortification experiments and the calculation of recovery values. An internal standard correction was performed by online addition of an internal standard solution of Rh and Y (50 μg L^{-1}) in a T-piece. Duplicate analysis was performed for each sample. The HNO$_3$ concentration of external standard solutions was accurately matched to the final concentration of HNO$_3$ in the samples (i.e., 3.8%).

2.4. ICP-MS Analysis of Elements

ICP-MS analyses were performed by using a 7500A ICP quadrupolar mass spectrometer (Agilent Technologies, Palo Alto, CA, USA) fitted with an ASX-510 autosampler (CETAC, Omaha, NE, USA) and a peristaltic pump. A Babington nebulizer with a Scott spray chamber (Agilent Technologies) was used for sample introduction. Detailed operating conditions and instrumental parameters are given in Table S2. The optimization of ICP-MS was carried out to obtain maximum signal intensities for 7Li, 89Y, 140Ce and 205Tl using a tuning solution while keeping the formation of oxides 140CeO$^+$/140Ce$^+$ and doubly charged species Ce^{2+}/Ce$^+$ ratios below 1% and 2%, respectively. Pollen and honey samples were analyzed for Mg, Al, K, Ca, V, Cr, Mn, Fe, Co, Ni, Zn, Cu, As, Rb, Sr, Cd, Cs, Tl, Pb and U. The external standard method was applied for quantification, after which we verified the accuracy of the method with fortification experiments and calculation of recovery values. An internal standard correction was performed by the online addition of an internal standard solution of Rh and Y (50 μg L^{-1}) in a T-piece. Duplicate analysis was performed for each sample. The HNO$_3$ concentration of external standard solutions was accurately matched to the final concentration of HNO$_3$ in the samples (i.e., 3.8%). Data analysis was performed using ChemStation software (version G1834B) (Agilent Technologies).

2.5. Risk Assessment

The non-carcinogenic and carcinogenic health risks through the consumption of honey and bee pollen were assessed according to the estimated daily intake (*EDI*), target hazard quotient (*THQ*), hazard index (*HI*) and lifetime cancer risk (*LTCR*) [48–50].

2.5.1. Non-Carcinogenic Risk

The *THQ* is the probable non-carcinogenic risk for orally ingested elements; it is defined as the ratio of the daily oral intake to the oral reference dose with the following equation, as suggested by the United States Environmental Protection Agency (US EPA):

$$THQ = \frac{EDI}{RfDm}$$

The estimate daily intake (*EDI*) value was calculated according to the formula suggested by USEPA and other authors [51–53].

$$EDI = (C \times IR \times EF \times TE)/(BW \times AT)$$

where *C* is the concentration of each potentially toxic element (PTE) detected in the samples (mg/kg), *IR* is the intake rate of honey and bee pollen (kg/day), *EF* is the exposure frequency to the contaminant (350 day/year), *TE* is the total exposure, and *AT* is the average lifetime time for non-carcinogenic risk (*TE* × 365 day/year). The dates related to *BW*, *AT*, *TE* and *IR* that are related to different groups and used for the assessment of *EDI* are reported in Table 1.

Table 1. Values of parameters used for the assessment of EDI.

Category	Years	Body Weight (BW) (kg)	Average Lifetime (AT) (Days)	Total Exposure (TE) (Years)	Intake Rate (IR) Honey [A] (kg/Day)	Intake Rate (IR) Bee Pollen [B] (kg/Day)
toddler	0–3	11.3	730	2	0.0127	-
children	3–10	26.1	2555	7	0.0126	0.02
adolescent	10–18	52.6	2920	8	0.0133	-
adult	18–65	69.7	17155	47	0.0127	0.04

Bibliographic references for values used in this study: [A] [2,54–56]; [B] [11,27,52,56–59].

RfD_m is the oral reference dose ($mg/kg_{bw}/day$) (Table 2). Given the challenges in setting a reliable threshold for lead (Pb) according to the USEPA, this study relied on the RfD_{Pb} (reference dose for lead) proposed by previous research as a suitable alternative [60–63].

Table 2. Oral reference dose ($mg/kg_{bw}/day$) and cancer slope factor $(mg/kgbw/day)^{-1}$ for each element.

Elements	RfD ($mg/kg_{bw}/Day$)	Reference	CSF ($mg/kg_{bw}/Day$)	Reference
		[53]		/
Cd	0.0001	[51]	0.38	[62,64,65]
V	0.005 **	[51]	/	/
Cr	0.003 ***	[51]	0.50	[66]
Ni	0.02 ▲	[51]	1.70 ○	[66]
Cu	0.04	[51]	/	/
As	0.0003 ▲▲	[51]	1.50	[51]
Ba	0.20	[51]	/	/
Sb	0.0004 ▲▲▲	[51]	/	/
Pb	0.0035	[60–63]	0.0085 ○○	[66]
Mn	0.10	[51]	/	/
Al	1	[67]	/	/
Fe	0.7	[68]	/	/
Co	0.0003	[67]	/	/
Rb	0.005	[69]	/	/
Zn	0.3	[67]	/	/
U	0.003	[70]	/	/

** Vanadium and Compounds, *** Chromium VI, ▲ Nickel Soluble Salts, ▲▲ Inorganic Arsenic, ▲▲▲ Antimony (metallic) and Antimony Tetroxide, ○ Nickel Subsulfide, ○○ Lead and Compounds. Adapted from [2].

A THQ_m (dimensionless) >1 entails a high non-carcinogenic risk, as the adverse health effect is considerable, while, if THQ_m is <1, it is generally presumed to be safe for the risk of non-carcinogenic effects.

The cumulative risk arising from the dietary exposure to all elements in the same foodstuff, in our case honey or bee pollen, was assessed through the Hazard Index (*HI*). representing the cumulative sum of THQ_m values for each element and calculated as follows:

$$HI = \sum_m THQm$$

A *HI* > 1 entails a high potential health impact implication, at the opposite a *HI* < 1 indicates that there is no apparent health impact due to the metals considered. A serious chronic health impact has been suggested for *HI* > 10 [2].

2.5.2. Carcinogenic Risk

The *LCTR* is the carcinogenic effect related to the ingestion of food contaminated by Ni, Cr, Pb, As, and Cd [51].

$$LTCR = EDI \times CSF$$

CSF represents the cancer slope factor $(\mathrm{mg/kg_{bw}/day})^{-1}$ that estimates the probability of developing cancer due to the ingestion of Ni, Cr, Pb, As, and Cd. The $\mathrm{CSF_{Cd}}$ proposed (Table 1) was previously used by other authors [62,64,66].

The US EPA considers an *LTCR* (dimensionless) $>1 \times 10^{-4}$ as an unacceptable risk in regard to developing cancer over a human lifetime. *LTCR* values between 1×10^{-6} and 1×10^{-4} are considered to be an acceptable range for carcinogenic risk. The Canadian Safe Environments Directorate (2010) proposes the value of 1×10^{-5} as the maximum safety threshold for the risk of developing cancer [71].

The cumulative cancer risk is the risk estimation due to exposure to multiple carcinogenic elements and is calculated as:

$$LTCRtot = \sum_{k=1}^{n} LTCRk$$

where *LTCRk* is the life time cancer risk for the cancer element *k*.

2.6. Statistical Analysis

Data were expressed as mean $\pm$ standard deviation. One way ANOVA and a Kruskal–Wallis test were used to investigate significant differences among samples where the a confidence level was held at 95%. Principal component analysis (PCA) and hierarchical cluster analysis (HCA) were performed with honey and bee pollen datasets. Data analysis was performed using XLSTAT software (version 2023.3.1) (Addinsoft SARL, New York, NY, USA) and ClustVis, a web tool freely available at http://biit.cs.ut.ee/clustvis/ (accessed on 1 February 2024) [72].

3. Results

3.1. Use of Soil and Characterization of the Flight Areas (CLC)

Bees, flying in their extensive foraging areas, come into contact with air, water, and soil, potentially picking up contaminants like PTEs and transferring to their hives and hive products. Therefore, investigating soil usage is crucial. The study examined four specific areas (Figure 2) based on information provided from the CLC project map, modified by ArcGis 10.6 software (Redlands, CA, USA), and significant differences among the specific uses of soil were highlighted. The Apiary 1 (A1) flight area was the one mainly characterized by the presence of a continuous (code 111) and discontinuous (code 112) urban fabric at around 11%, followed by complex cultivation and vineyards at around 76%. The Apiary 2 (A2) flight area was represented mainly by the presence of agriculture, with significant areas of natural vegetation (code 243) and complex cultivation patterns (code 242) at around 83% and a small portion of discontinuous urban fabric and vineyards. The Apiary 3 (A3) flight area was covered mainly by vineyards ($\approx$57%), a portion of non-irrigated arable land ($\approx$25%), generally under a crop rotation system, and a smaller area of discontinuous urban fabric and complex cultivation patterns ($\approx$18%). The Apiary 4 (A4) flight area was characterized by the significant presence of vegetation formation composed principally of trees (code 311) at around 32%, followed by non-irrigated arable land ($\approx$45%), complex cultivation systems ($\approx$10%), and agro-forestry and natural vegetation areas at around 8%. In brief, the A1 apiary was situated in the most heavily anthropic environment, whereas the A4 flight area was positioned within the Natural Regional Reserve of Castel Cerreto (Teramo, Abruzzo), representing the least anthropized environment. A2 and A3 exhibited intermediate levels of anthropization.

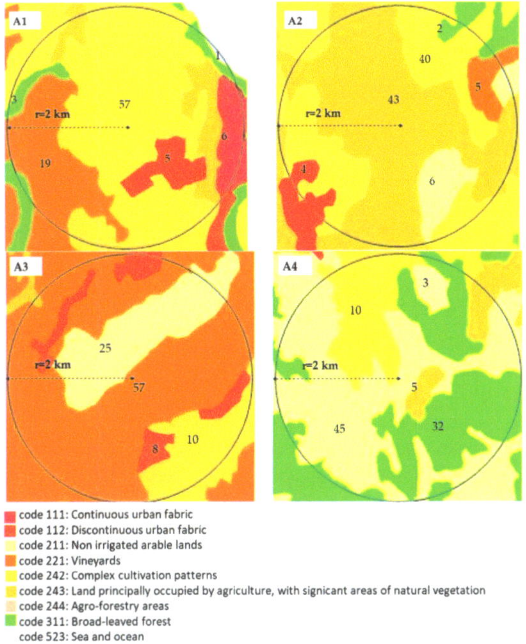

Figure 2. Corine Land Cover use in the beehive areas (r = 2 km) of A1: Ortona (CH), A2: Filetto (CH), A3: Frisa (CH), A4: S. Andrea (TE). Numbers in figure represent the percentage of use by the different types of area, as coded.

3.2. Mineral Contents of Honey and Bee Pollen Samples

The complex interplay of bees' environment, vegetation, floral sources, climate, and geographical traits gives rise to unique varieties of honey. The resulting mineral profile serves as a crucial tool for evaluating its nutritional value, identifying its geographic origin, and detecting environmental contamination by heavy metals [6]. The concentrations of the twenty elements detected in multifloral honey samples obtained from the four different areas are reported in Table 3. Except for the A1 area, all the other samples were harvested both in 2019 and 2020.

Table 3. Elemental composition of honeys ($\mu g\ g^{-1}$ ± standard deviation (n = 3)).

Element	A1 2020	A2 2019	A2 2020	A3 2019	A3 2020	A4 2019	A4 2019	A4 2020
Mg	10.875 ± 0.339	20.016 ± 1.163	29.496 ± 0.509	15.703 ± 0.940	19.547 ± 0.634	31.22 ± 0.849	9.584 ± 0.387	25.438 ± 0.175
Al	0.351 ± 0.042	n.d	2.314 ± 0.251	1.485 ± 0.066	n.d.	0.607 ± 0.021	0.117 ± 0.017	0.188 ± 0.008
K	445.962 ± 2.448	542.353 ± 20.797	885.447 ± 27.102	547.545 ± 6.164	532.569 ± 23.354	797.00 ± 3.030	423.783 ± 0.553	549.793 ± 5.435
Ca	22.155 ± 1.091	35.103 ± 1.269	48.314 ± 0.591	28.199 ± 1.977	34.665 ± 0.525	69.32 ± 0.465	15.217 ± 0.317	63.533 ± 0.964
V	0.001 ± 0.000	0.001 ± 0.000	0.005 ± 0.000	0.003 ± 0.000	0.001 ± 0.001	0.001 ± 0.000	n.d	0.001 ± 0.000
Cr	0.238 ± 0.008	0.243 ± 0.008	0.234 ± 0.004	0.238 ± 0.006	0.235 ± 0.008	0.205 ± 0.003	0.203 ± 0.001	0.176 ± 0.004
Mn	0.230 ± 0.008	0.131 ± 0.007	0.412 ± 0.011	0.388 ± 0.013	0.126 ± 0.009	0.261 ± 0.000	0.085 ± 0.001	0.320 ± 0.002
Fe	n.d	n.d	n.d	n.d	n.d	n.d	n.d	n.d
Co	0.001 ± 0.000	0.002 ± 0.000	0.004 ± 0.000	0.003 ± 0.000	0.002 ± 0.000	0.003 ± 0.000	0.001 ± 0.000	0.003 ± 0.000
Ni	0.019 ± 0.000	0.015 ± 0.002	0.038 ± 0.004	0.021 ± 0.003	0.017 ± 0.001	0.019 ± 0.001	0.026 ± 0.000	0.031 ± 0.000
Cu	0.206 ± 0.003	0.173 ± 0.010	0.564 ± 0.009	0.361 ± 0.019	0.167 ± 0.007	0.209 ± 0.005	0.091 ± 0.000	0.222 ± 0.005
Zn	0.157 ± 0.002	0.364 ± 0.046	0.778 ± 0.023	0.373 ± 0.056	0.639 ± 0.014	2.956 ± 0.095	0.602 ± 0.009	2.826 ± 0.067
As	0.003 ± 0.001	0.006 ± 0.001	0.006 ± 0.001	0.004 ± 0.000	0.005 ± 0.001	0.004 ± 0.001	0.002 ± 0.001	0.003 ± 0.000
Rb	0.499 ± 0.003	0.429 ± 0.017	0.567 ± 0.020	0.412 ± 0.027	0.408 ± 0.015	0.415 ± 0.027	0.175 ± 0.003	0.349 ± 0.004
Sr	0.081 ± 0.001	0.131 ± 0.008	0.205 ± 0.001	0.116 ± 0.009	0.131 ± 0.001	0.356 ± 0.005	0.086 ± 0.002	0.367 ± 0.006
Cd	n.d	0.0003 ± 0.0002	n.d	n.d	n.d	n.d	n.d	n.d
Cs	0.000 ± 0.000	0.001 ± 0.000	0.001 ± 0.000	0.000 ± 0.000	0.001 ± 0.000	0.000 ± 0.000	n.d	0.000 ± 0.000
Tl	0.001 ± 0.000	0.002 ± 0.000	0.001 ± 0.000	0.001 ± 0.000	0.001 ± 0.000	0.001 ± 0.000	0.001 ± 0.000	0.000 ± 0.000
Pb	0.039 ± 0.001	0.064 ± 0.003	0.002 ± 0.000	0.032 ± 0.002	0.052 ± 0.002	0.009 ± 0.001	0.001 ± 0.001	0.013 ± 0.000
U	0.000 ± 0.000	0.001 ± 0.000	0.000 ± 0.000	0.000 ± 0.000	0.001 ± 0.000	0.000 ± 0.000	0.000 ± 0.000	0.000 ± 0.000

n.d: not detected.

Concerning the essential elements, the most abundant macrominerals were K, Ca, and Mg, with mean values of 590.6, 39.6 and 20.2 µg g^{-1}, respectively. Despite the great variability, all the element results were in line with the content observed by different authors for honeys from the center and south of Italy, indicating K to be the most abundant mineral in honey, followed by Ca, Na, and Mg [6,18,73–76]. Conversely, iron content was not detected (<LOD) in honey samples.

Trace elements such as Mn, Co, Cu, Zn, Rb, Sr, Al, and Tl showed mean values of 0.244, 0.002, 0.249, 1.086, 0.407, 0.184, 1.11, and 0.0009 µg g^{-1}, respectively. Among these, either Zn, in sample A4 (with a mean value of 2.13 µg g^{-1}), and Al, in sites A2 and A3 (with means values of 2.31 and 1.49 µg g^{-1}, respectively), showed the highest values.

With regard to heavy metals, the mean concentrations were of 0.004 µg g^{-1} (As), 0.0231 µg g^{-1} (Ni), 0.0317 µg g^{-1} (Pb) and 0.222 µg g^{-1} (Cr), while Cd was not detected (<LOD) in the samples. Interestingly, the levels of Pb and Cr were significantly higher ($p < 0.05$) in A1, A2, and A3 with respect to A4, denoting an anthropic pollution in the first three areas, as observed previously. Indeed, Cr has been reported to be very widespread in the environment, and, in absence of metallurgical and chemical manufacturing industries located near the hives, it could be transferred to different distances due to the wind action, meteorological factors, topography, and vegetation, which are strictly related to the long-transfer of the metal [35]. Pb is one of the most widespread environmental pollutants, and this is mainly attributed to internal-combustion engines [77]. Regarding the high affinity of Pb as an atmospheric particular matter, the presence of emission sources of particles like road asphalt and tires around the hive can cause honey to be contaminated with Pb. All the honey samples resulted within the legal limits for lead content (100 µg kg^{-1}), considering that the highest level of Pb in honey samples was 53.5 µg kg^{-1}, confirming the excellent quality of the analyzed honeys and the lower transfer capacity of the elements from the environment via bees to the final product. No significant differences were highlighted for nickel content ($p < 0.05$), while higher values ($p > 0.05$) were found for arsenic in the A2 and A3 areas, with mean values of 0.005 and 0.004 µg g^{-1} respectively.

The element contents found in this study aligned with the literature data previously obtained by other authors and reported in Table S3. However, it is possible to highlight the great variability in terms of qualitative and quantitative composition due to biotic and abiotic factors, such as the effect of anthropic pollution frequently reported by other authors, even within the same variety [78,79]. Furthermore, some research considered different or fewer elements than those observed in the present study. The values we found in this investigation were generally comparable to the values of honey originating from different Italian areas [40,80,81]. Furthermore, differences in analytical approach, including the methods of sample solubilization and determination techniques, may also affect the results [82].

The elemental composition of bee pollen is reported in Table 4. No significant differences ($p < 0.05$) were observed for Ca, V, Cr, Cu, As, Cd, Cs, Tl, and U among the four areas and between the two harvesting periods (July and October), with mean values of 1223, 0.055, 0.187, 12.95, 0.054, 0.040, 0.038, 0.032 and 0.031 µg g^{-1}, respectively. Concerning the essential elements, despite significant differences among the samples ($p < 0.05$), the most abundant macrominerals were K, Ca, and Mg, with mean values of 5985, 1233, and 853 µg g^{-1}, respectively. The same behavior was observed for the non-essential minerals, where the harvest area significantly influenced the elemental content of bee pollen ($p < 0.05$). This peculiar aspect highlights the difficult-to-compare literature data of bee pollen from both different Italian regions and foreign countries; therefore, for the sake of clarity, an exhaustive summary of bee pollen mineral composition was reported in Table S4. Overall, the results found in the present study are in accordance with the content reported by different authors [13,27,48,58,73,74,83].

Table 4. Elemental composition of polyfloral bee pollen ($\mu g\ g^{-1}$ ± standard deviation (n = 3)).

Element	A1	A1	A2	A2	A3	A3	A4	A4	A4
Mg	976 ± 104	798 ± 60	621 ± 1	819 ± 6	1089 ± 36	813 ± 35	857 ± 58	962 ± 47	751 ± 55
Al	4 ± 1	25 ± 2	6 ± 1	9 ± 1	6 ± 1	6 ± 1	8 ± 1	31 ± 1	18 ± 1
K	5185 ± 412	5711 ± 399	5930 ± 23	5787 ± 135	5929 ± 82	5543 ± 143	6370 ± 453	7186 ± 261	6229 ± 273
Ca	1328 ± 74	1066 ± 87	1001 ± 8	1212 ± 71	1591 ± 38	1067 ± 24	1059 ± 51	1549 ± 71	1231 ± 48
V	0.08 ± 0.02	0.07 ± 0.01	0.06 ± 0.05	0.04 ± 0.01	0.04 ± 0.03	0.03 ± 0.01	0.04 ± 0.01	0.09 ± 0.01	0.06 ± 0.01
Cr	0.18 ± 0.04	0.20 ± 0.01	0.18 ± 0.06	0.11 ± 0.02	0.23 ± 0.04	0.15 ± 0.02	0.17 ± 0.01	0.28 ± 0.01	0.21 ± 0.01
Mn	71 ± 3	24 ± 1	23 ± 1	32 ± 2	87 ± 4	27 ± 0.4	24 ± 1	28 ± 1	20 ± 1
Fe	60 ± 2	56 ± 4	82 ± 8	39 ± 2	78 ± 4	29 ± 1	76 ± 2	113 ± 3	120 ± 11
Co	0.10 ± 0.07	0.08 ± 0.01	0.06 ± 0.05	0.03 ± 0.01	0.08 ± 0.03	0.03 ± 0.01	0.03 ± 0.01	0.05 ± 0.004	0.04 ± 0.003
Ni	1.89 ± 0.01	1.07 ± 0.04	0.88 ± 0.09	0.14 ± 0.02	2.50 ± 0.20	0.18 ± 0.02	1.31 ± 0.01	1.96 ± 0.02	1.40 ± 0.10
Cu	12.3 ± 0.5	14.1 ± 0.7	11.7 ± 0.6	13.6 ± 0.8	14.1 ± 0.8	11.5 ± 0.2	11.6 ± 0.2	16.0 ± 0.2	11.8 ± 0.2
Zn	47.0 ± 0.4	51 ± 1	44 ± 3	65 ± 4	57 ± 3	53 ± 1	179 ± 8	109 ± 4	76 ± 2
As	0.09 ± 0.10	0.05 ± 0.02	0.07 ± 0.07	0.04 ± 0.02	0.05 ± 0.05	0.05 ± 0.02	0.03 ± 0.01	0.05 ± 0.01	0.04 ± 0.001
Rb	7.71 ± 0.07	9.70 ± 0.50	6.90 ± 0.70	2.60 ± 0.20	4.90 ± 0.40	3.00 ± 0.10	5.07 ± 0.10	4.90 ± 0.03	4.42 ± 0.04
Sr	1.90 ± 0.20	1.22 ± 0.001	2.10 ± 0.20	2.10 ± 0.10	2.60 ± 0.20	1.70 ± 0.09	4.87 ± 0.04	6.55 ± 0.01	5.78 ± 0.06
Cd	0.08 ± 0.10	0.04 ± 0.01	0.05 ± 0.06	0.04 ± 0.02	0.04 ± 0.04	0.03 ± 0.02	0.02 ± 0.01	0.01 ± 0.01	0.01 ± 0.003
Cs	0.08 ± 0.09	0.05 ± 0.01	0.06 ± 0.06	0.02 ± 0.01	0.04 ± 0.04	0.02 ± 0.02	0.02 ± 0.009	0.02 ± 0.01	0.02 ± 0.002
Tl	0.08 ± 0.09	0.02 ± 0.01	0.05 ± 0.06	0.02 ± 0.02	0.04 ± 0.04	0.02 ± 0.02	0.01 ± 0.01	0.01 ± 0.01	0.01 ± 0.003
Pb	0.40 ± 0.10	13.80 ± 0.70	1.40 ± 0.20	0.66 ± 0.06	1.70 ± 0.20	0.64 ± 0.04	0.05 ± 0.01	0.06 ± 0.01	0.05 ± 0.02
U	0.08 ± 0.09	0.02 ± 0.01	0.05 ± 0.06	0.02 ± 0.02	0.03 ± 0.04	0.02 ± 0.02	0.01 ± 0.010	0.01 ± 0.007	0.01 ± 0.003

Focusing on heavy metals, no significant differences ($p < 0.05$) were observed for Cr, As, and Cd, with mean values of 0.187, 0.054, and 0.040 $\mu g\ g^{-1}$. Conversely, Ni and Pb highlighted significant differences ($p < 0.05$) among areas. Nickel had the lowest and the highest results, as seen in A2 (0.51 $\mu g\ g^{-1}$) and A4 (1.56 $\mu g\ g^{-1}$), respectively. Lead resulted significantly higher ($p < 0.05$) in the A1 area (7.1 $\mu g\ g^{-1}$) than in A2, A3, and A4, where no differences were denoted (1.03, 1.19 and 0.06 $\mu g\ g^{-1}$). Several studies support the toxic metal concentrations of bee pollen significantly depending on the degree of environmental pollution [11,32,84–86]. In the case of Ni, particularly high in A4, A1, and A3 ($p > 0.05$), it could be influenced by the natural geochemistry of soils [87,88], industrial processes, vehicle emissions, the combustion of fossil fuels, waste disposal, or the use of pesticides in agricultural practices [89].

Comparing the mean mineral content in bee pollen and honey (Figure 3), it was possible to highlight that the elemental profile exhibited similar geospatial trends or patterns associated with the same origin of the two beehive products. The detected honey concentration in the decreasing range (mean concentrations) was K > Ca > Mg > Zn > Fe > Mn > Cu > Al > Rb > Sr > Pb > Ni > Cr > Co > V>As > Cs > Tl > U, while in bee pollen it was K > Ca > Mg > Al > Zn > Rb > Cu > Mn > Cr > Sr > Pb > Ni > As > Co > V>Tl > Cs > U. Overall, the concentrations of bee pollens were 10–70 times higher compared to that in honey, similar to what has been observed by other authors, supporting the hypothesis of biological reduction in the levels of metals in the finished product. Indeed, this aspect is associated with the activity of bee enzymes during the honey elaboration process or with the presence of molecules such as gluconic and ascorbic acid, responsible for the chelation of elements and complex formations, leading to the absorption and accumulation of metals in specific body anatomic sections or excretion with feces rather than their accumulation in honey [90,91]. Conversely, differences in the pattern were denoted for Cr and Mn that showed a smaller (0.82) and a higher ratio (153), respectively.

As frequently mentioned, the mineral content of bee pollen and honey is strictly related to the vegetal species in terms of metabolism, physiology, and morphology, which influence the amount of elements in the different parts of the plant tissues, flowers included [89]. It was also observed that the capability of certain plants to concentrate pollutants can also affect their concentration in the honey sample. For example, honey obtained from the nectar of aromatic plants is characterized by a high concentration of heavy metals since they tend to concentrate pollutants more than herbaceous plants [75]. Furthermore, polluted bee pollen results in higher levels of metals than in honey, suggesting the potential use of such products as indicators of metal pollution in their areas of origin as well as of potential health risks [16].

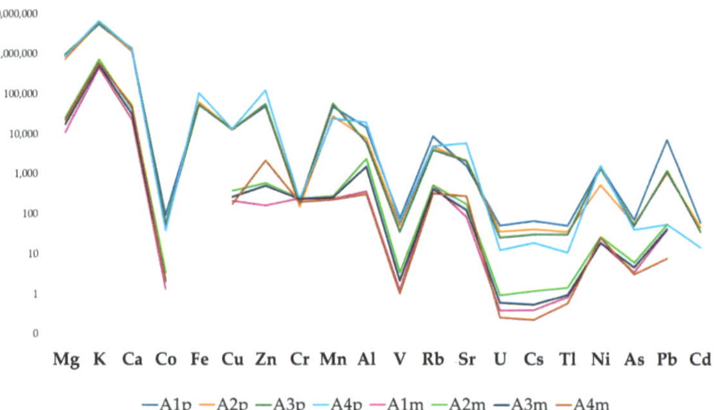

Figure 3. Elemental pattern of bee pollen (p) and honey (h) samples in the respective apiary (A1–A4).

3.3. Carcinogenic and Non-Carcinogenic Effects in Bee Pollen and Honey

The bioaccumulation of PTEs in a body fed by plants, feeds and animal-origin foods, as well as water, can contribute to a wide variety of adverse health effects, including organ damage, developmental alterations, and cancer [92]. Specific regulations regarding the presence of PTEs in honey and bee pollen are currently lacking. However, the Codex Alimentarius includes a stipulation that honey must be devoid of quantities of metals that could pose a hazard to human health.

In the present study, PTE accumulation rates and possible risk levels were estimated according to the daily honey and bee pollen consumption amount. For honey, more frequently consumed than bee pollen, the recommended daily dose is at around 10 g for toddlers, adolescents, and adults. In the case of bee pollen, the main consumers follow a health and environmentally conscious lifestyle, as well as the elderly, who use it due to its antioxidant and other therapeutic effects. Its recommended daily dose consumption was reported to range from 20 to 40 g for children and adults, respectively.

3.3.1. Non-Carcinogenic Risk (EDI, THQ_m, HI)

The EDI estimates the daily exposure level of the human population to toxic and potentially toxic elements through food consumption. The mean estimated daily intake (EDI) of the analyzed metals were assessed for toddlers, children, adolescents, and adults for honey, as well as for bee pollen in regard to children and adults, due to the poor information about bee pollen consumption in these categories. The total EDI rank of all metals for honey follows the decreasing order of toddlers (0.69 mg/day) > children (0.29 mg/day) > adolescents (0.15 mg/day) > adults (0.11 mg/day), while, for bee pollen, the trend is children (6.03 mg/day) > adult (4.52 mg/day).

The honey and bee pollen EDI ranks of individual metals for all groups follow the decreasing order of K > Ca > Mg > Al > Zn > Rb > Cu > Mn > Cr > Sr > Pb > Ni > As > Co > V>Tl > U, and K > Ca > Mg > Zn > Fe > Mn > Cu > Al > Rb > Sr > Pb > Ni > Cr > Co > V>As > Cd > Cs > Tl > U, respectively. Interestingly, the EDI related to the apiaries shows the rank order A2 > A4 > A3 > A1 for honey and A3 > A4 > A2 > A1 for pollen.

Overall, the EDI of each metal obtained is reported to be lower than the correspondent maximum tolerable daily intake for both honey and pollen.

The THQ_m values for honey and bee pollen were reported in Figures 4 and 5, respectively. For all the analyzed elements, the THQ_m values in honey were below 1, suggesting that the exposed human population is supposed to be safe [48]. Conversely, in bee pollen samples, and mainly for those belonging to the A1 apiary, the THQ_m value resulted above 1 for Pb for both the children and adult group, indicating a potential health risk associated with its consumption.

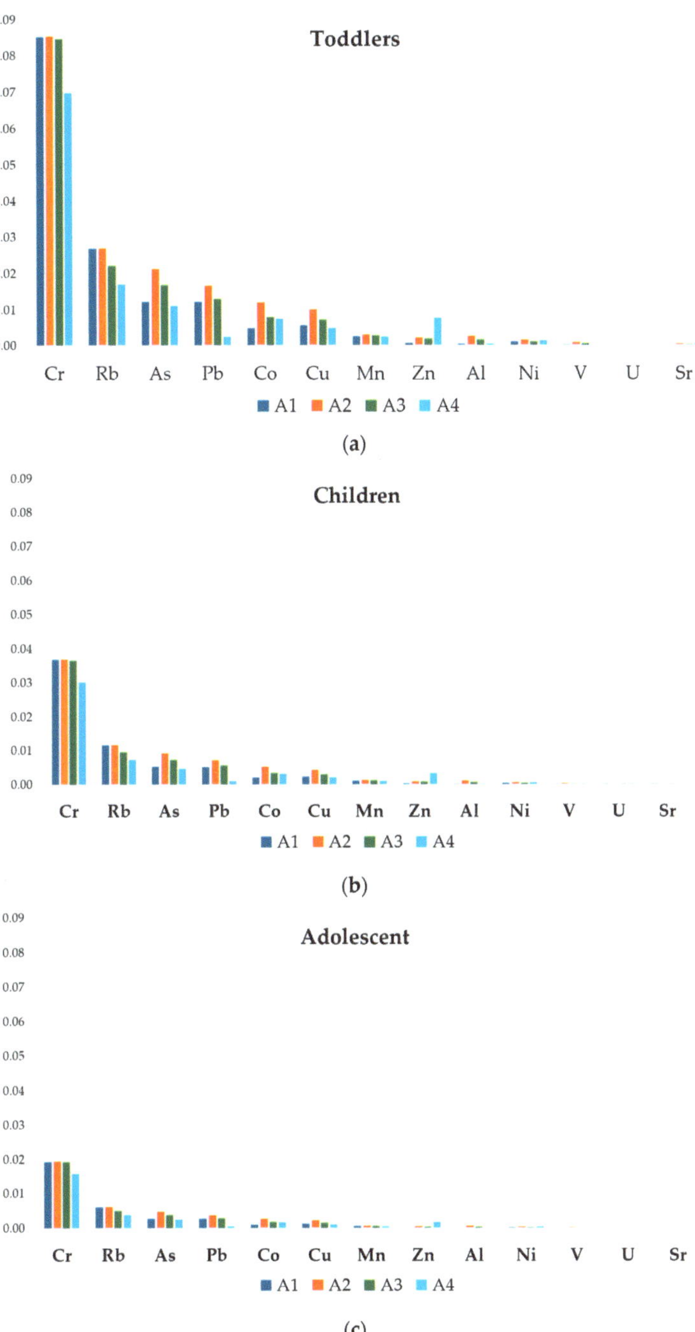

Figure 4. *Cont.*

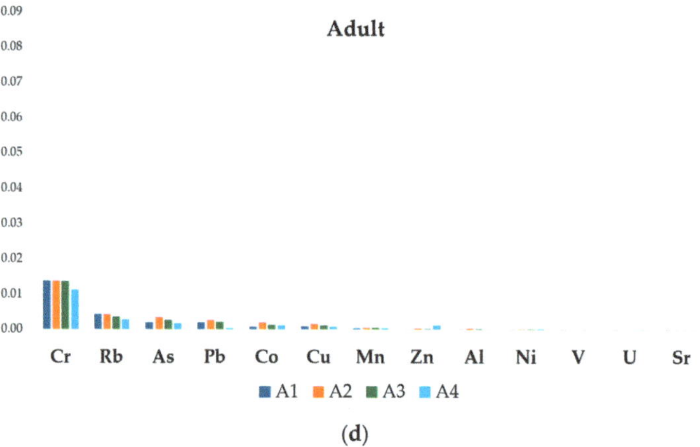

(**d**)

Figure 4. *THQ$_m$* value for honey in toddlers (**a**), children (**b**), adolescent (**c**) and adult (**d**) in different apiaries.

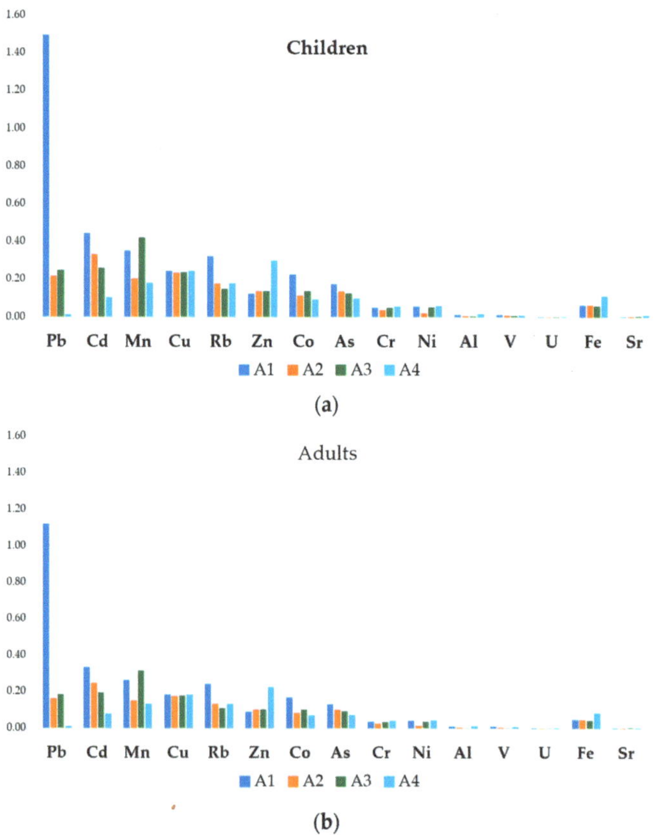

(**a**)

(**b**)

Figure 5. *THQ$_m$* value for bee pollen in children (**a**) and adult (**b**).

In the four apiaries, the honey average THQ_m exposure values ranged from 8.5×10^{-2} (Cr) in toddlers (A1) to 1.4×10^{-6} (U) in adults (A4), while in bee pollen, the THQ_m higher values ranged from 1.5 (Pb) in children (A1) to 2.2×10^{-4} (U) in adults (A4).

Considering all groups, the honey and bee pollen THQ_m values of individual metals followed the decreasing order of Cr > Rb > As > Pb > Co > Cu > Zn > Mn > Ni > Al > V>U, and Pb > Cd > Cu > Mn > Rb > Zn > Co > As > Cr > Ni > Al > V>U, respectively.

The sum of the THQ_m values for each category, represented by the *HI* index, was reported in Figure 6. Honey samples (Figure 6a) showed values below the safety threshold (<1) for all consumer groups and apiaries, resulting in no health concerns. Contrarily, for bee pollen samples (Figure 6b), only the adult group in the A4 apiary presented a *HI* value below 1; therefore, the consumption of bee pollen belonging to the A1, A2, and A3 apiaries represented a health risk concern. For honey, the average *HI* risk rank, based on the consumer groups, was toddlers (0.154) > children (0.066) > adolescents (0.035) > adults (0.025), while for bee pollen it was children (2.1) > adults (1.5), higher than honey because of the greater metal concentrations.

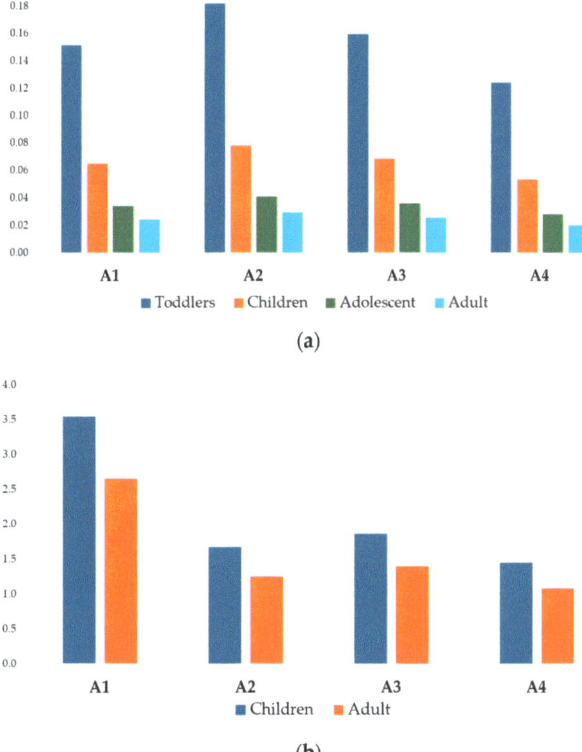

Figure 6. *HI* index for honey (**a**) and bee pollen (**b**) for different consumer categories and apiaries.

The honey average *HI* rank order based on all apiaries corresponded to A2 > A3 > A1 > A4, while in bee pollen it was A1 > A3 > A2 > A4. For honey, the highest HI value occurred for toddlers in the A2 apiary (0.18), and it was the lowest for the adult group in the A4 apiary (0.020); for bee pollen, the highest *HI* value occurred for children in the A1 apiary (3.5) and the lowest occurred for the adult group in the A4 apiary (0.99). It was quite evident that the A4 apiary could be considered safer in terms of honey and bee pollen than the other sites.

The average percentual contribution of PTEs, reported in Figure 7, highlighted that, in the case of honey consumption (Figure 7a), the 52.9% was accounted by Cr, followed by Rb (15%), As (9.9%), and Pb (7.1%), while the rest of the metals cumulatively accounted for only 15.1%. In the case of bee pollen consumption (Figure 7b), the main contribution was related to Pb (23.9%), followed by Cd (14%), Cr (13.8%), Mn (11.6%), Rb (9.9%), Zn (8.4%), Co (6.8%), As (6.4%), while the rest of the metals cumulatively accounted for 5.24%.

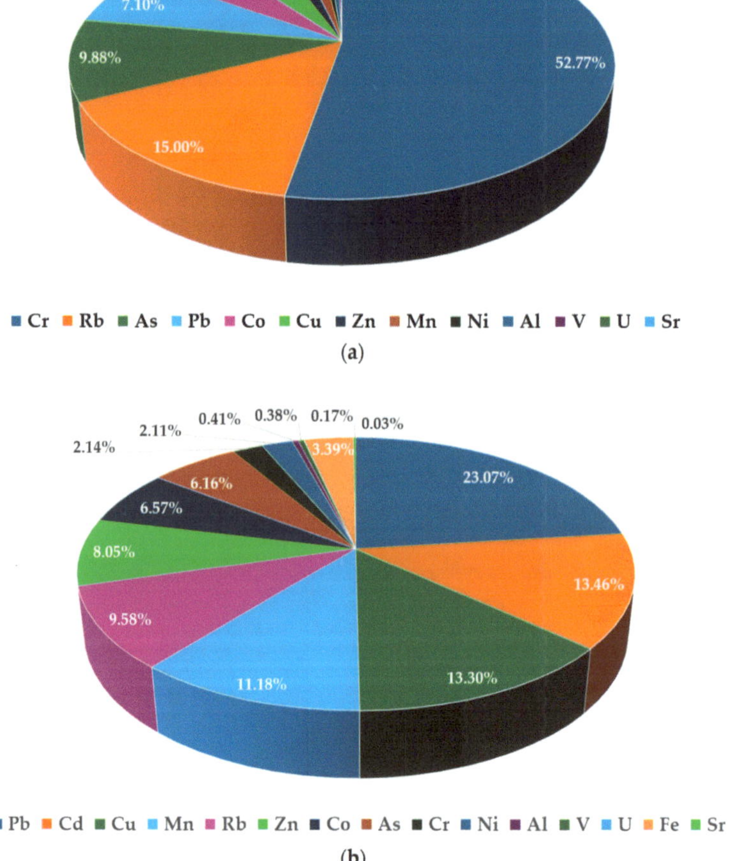

Figure 7. Contribution (%) of each metal to *HI* due to the consumption of honey (**a**) and bee pollen (**b**).

3.3.2. Carcinogenic Risk (LCTR)

The carcinogenic risk assessment (*LCTR*), calculated based on CSF values reported in Table 2, and particularly for Ni, Cr, Pb, As, and Cd, is shown in Figure 8. Concerning honey, *LTCR* value >1 × 10^{-4} was reported for Cr exposure in the toddlers category (Figure 8a) while *LTCR* value >1 × 10^{-5} was observed for children, adolescents, and adults (Figure 8b–d), following the apiary rank A2 > A1 > A3 > A4. Nickel *LTCR* values >1 × 10^{-5} were observed in toddlers and children, regardless the apiary, following the order A2 > A4 > A3 > A1, and for the adolescent category in apiaries A2 and A4.

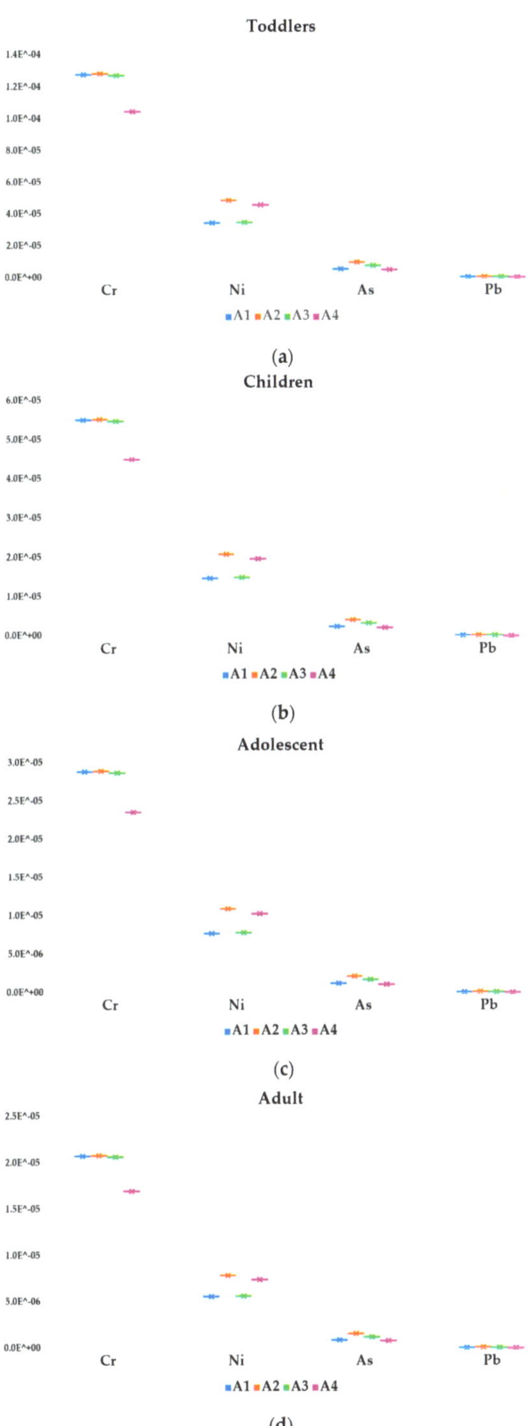

Figure 8. Lifetime cancer risk (*LCTR*) values based on carcinogenic elements exposure in toddlers (**a**), children (**b**), adolescent (**c**) and adults (**d**) in different apiaries.

The *LTCR* related to bee pollen is reported in Figure 9. A Ni *LTCR* value $>1 \times 10^{-4}$ was observed for both children and adults, and for all the apiaries in the following order A4 > A1 > A3 > A2, *LTCR* $>1 \times 10^{-5}$ was observed for Cr, As, Cd, and Pb despite slightly different apiary ranks. In particular, Cr followed A4 > A1 = A3 > A2, while As followed an A1 > A2 > A3 > A4 order. The *LTCR* value for Cd was $> 1 \times 10^{-5}$, observed in children and adults with an apiary rank of A1 > A2, while *LTCR* $>1 \times 10^{-5}$ was recorded for Pb in adults, especially in the A2 apiary.

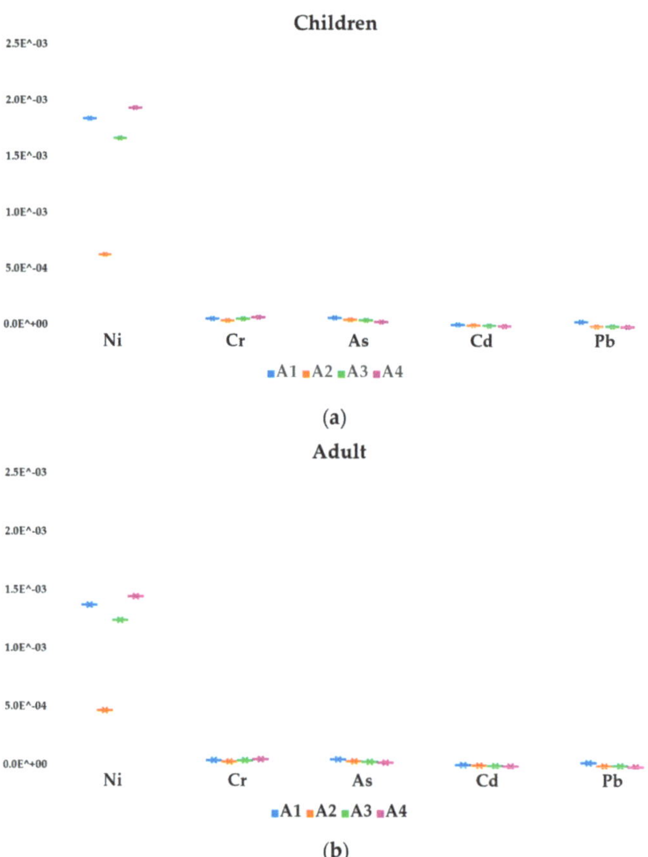

Figure 9. Lifetime cancer risk (LCTR) values based on carcinogenic element exposure in children (**a**) and adults (**b**) in different apiaries.

Several studies which characterized honey by values of *LCTR* included amounts of between 1×10^{-5} and 1×10^{-4}, like those reported in this study [63,77,93–96].

In the case of bee pollen, *LCTRs* above 1×10^{-5} and 1×10^{-4} are described by [27,56].

The contribution of each element to *LCTRs* in honey and bee pollen is reported in Figure S1. Cr accounted for 71.8%, followed by Ni (24%), As (4%), and Pb (0.2%) in honey samples (Figure S1a). Concerning bee pollen (Figure S1b), Ni accounted for 90.9%, then Cr (4.1%), As (3.5%), Pb (4.1%), and Cd (9.6%).

The cumulative cancer risk (*LCTRtot*) is reported in Figure 10. With regard to honey, due to exposure to multiple carcinogenic elements, *LCTRtot* was $>1 \times 10^{-4}$ in the case of toddlers, while it was $>1 \times 10^{-5}$ for children, adolescents, and adults. Conversely, the *LTCRtot* value for bee pollen highlights values ranging from 1.3×10^{-3} to 7.7×10^{-4}. The

LCTRtot rank based on apiaries corresponds to A2 > A3 > A1 > A4 for honey, while for bee pollen it is A4 > A1 > A2 > A3.

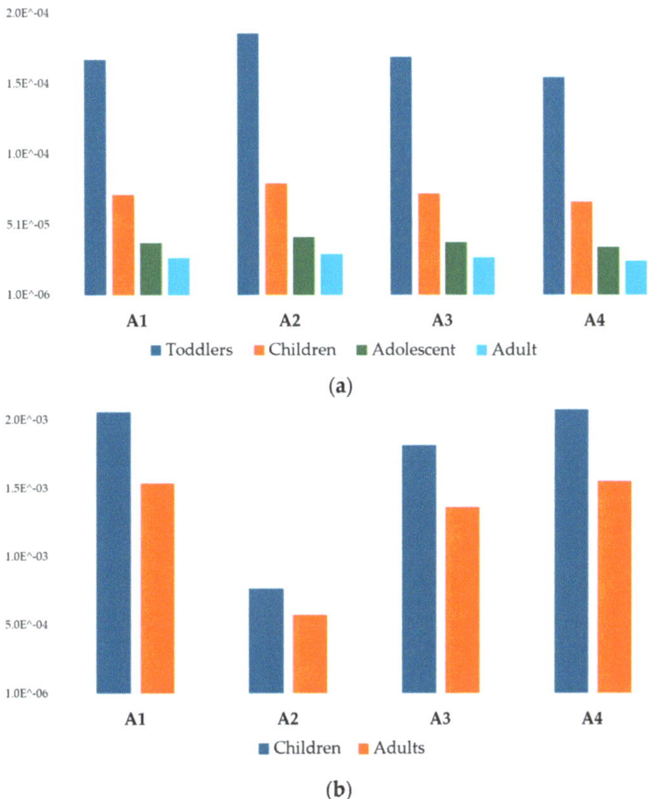

Figure 10. Cumulative lifetime cancer risk (LCTRtot) values based on carcinogenic element exposure in different consumer categories in honey (**a**) and bee pollen (**b**) in different apiaries.

3.4. PCA

The element content of honey and bee pollen samples were examined by PCA. Fe and Cd, non-detected in honey, were not considered. The biplot of loadings (variables) and score (observations), reported in Figure 11, highlighted a clear separation of the two macro samples, honey and bee pollen, along F1, which explains the 76.61% of the total variance (85.98%). Conversely, apiaries, regardless of the product, were well separated along the F2 component, since A1, A2 and A3 were located in the positive quadrants while A4 was in the opposite negative side. Further, bee pollens of A1, A2 and A3 were strongly correlated with most of the metal, except Cr, which was correlated with the honey belonging to the same apiaries. Observing the F2 component, apiary A4 was completely separated both for honey and bee pollen. In particular, bee pollen was found to be richer in Zn, opposite to A1, A2 and A3, which were richer in Pr, Tl, U, and As.

Results presented by PCA elaboration confirm those previously discussed; indeed, the products belonging to the apiaries A1, A2 and A3, unlike the A4 apiary, proved to be more contaminated with heavy metals such as Pb, Cr, As, and also Cd.

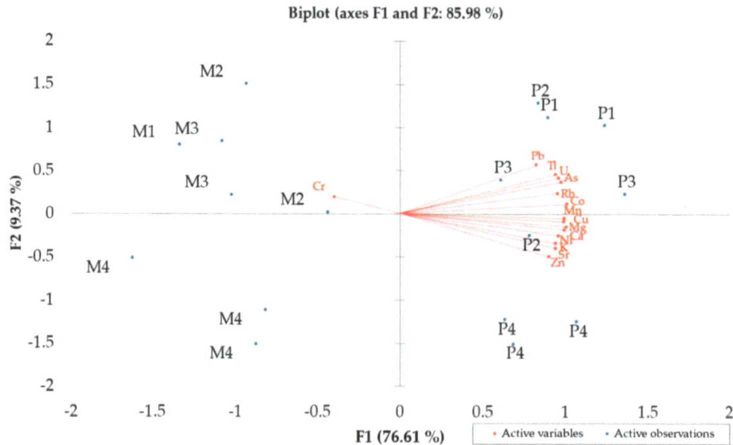

Figure 11. Principal component analysis (PCA) biplot showing the differentiation of the two bee product matrices by the first two principal axes.

3.5. HCA

An aggregative hierarchical cluster analysis (HCA), using Euclidean distances and Ward's linkage method, was implemented to obtain further data interpretations based on an input matrix consisting of 15 chemical variables (metals) and 17 samples among bee pollen and honey. The results of HCA for honey and bee pollen are shown in the heatmap plot (Figure 12).

Observing the honey HCA (Figure 12a), and in particular the rows, it was possible to highlight three metal groupings, with the first featuring the main heavy metals (Cr, As, Pb, Tl and U), the second being characterized with the macrominerals and some microelements (Ca, Mg, K, Zn, Sr), and the third having Ni, Rb, Mn, Co and Cu. Analysing honey samples, following such groupings, the high content of heavy metals in both the A2 and A3 samples was quite appreciable, while the A4 samples were found to be richer in macrominerals, which are important from a nutritional point of view.

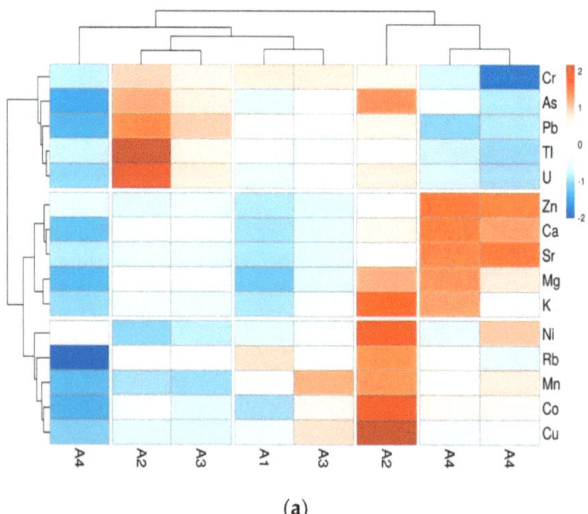

(a)

Figure 12. *Cont.*

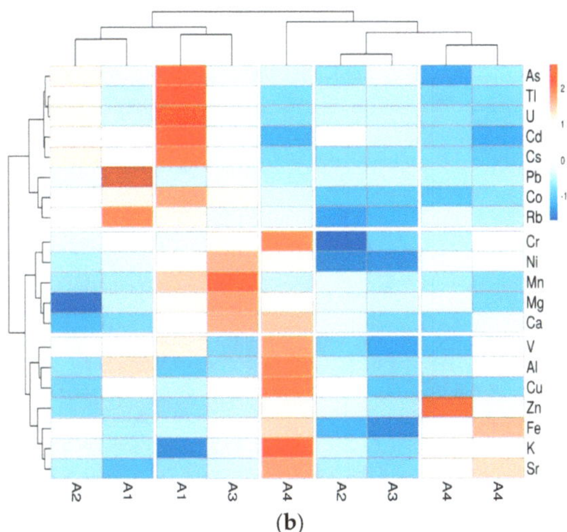

(b)

Figure 12. Hierarchical cluster analysis (HCA) of honey (**a**) and bee pollen (**b**) samples.

The results for bee pollen HCA were different (Figure 12b). By the row grouping, the first cluster was related to heavy metal except for Cr and Ni, which instead were grouped in the second cluster, followed by the rest of the metals being grouped in the third one. Both the A1 and A2 samples were grouped in regard to heavy metal content, with A1 in particular showing the highest levels of Pb, As and Cd; contrarily, A4 samples proved to poor in terms of the latest elements, except for the presence of Ni and Cr, confirming the wider mobility of bees, and especially in case of flowers scarcity related to meteorological or other adverse conditions giving access to areas wider than 50 km^2 and therefore coming into contact with more polluted areas [36,97].

The results presented confirmed those previously observed, mainly by PCA analysis. Further, it was quite difficult to directly correlate the presence of metals in bee pollen and honey due to the differences between the two products. Indeed, as frequently mentioned, numerous factors affect the content of metals in beehive products. In particular, it was evident that the effects of bee biotransformation of honey presented a more homogeneous grouping among the samples belonging to the same area than that of bee pollen.

4. Conclusions

In the frame of food safety, the multi-elemental profile of honeys and bee pollen provided information regarding both nutritional values and environmental conditions of the harvesting areas of the Abruzzo region. Results reveal differences in the mineral and metal content associated with the influence of biotic and abiotic factors characteristics of each specific area. Negligible values were found for potentially toxic metals such as cadmium, arsenic and lead, which were recovered at concentrations lower than the maximum limit set by European regulations.

Honey can be considered safe for consumption by adults, adolescents and children due to the low carcinogenic and non-carcinogenic risk values. However, there is particular concern for toddlers due to its high LCTRtot value, mainly associated with the accumulation of chromium (Cr) in the product. Despite the high nutritional value, bee pollen exhibited elevated LCTRtot levels in both the adult and children categories, primarily due to the accumulation of lead (Pb) and nickel (Ni), a particular note for attention in regard to for public health. The results also highlighted the relationship between the flight area, well described by the Corine Landcover maps, and the nutritional and safety properties of honey and bee pollen, indicating that the A4 apiary had better results because it was less anthropized.

Supplementary Materials: The following supporting information can be downloaded at: https://www. mdpi.com/article/10.3390/foods13121930/s1, Figure S1a: Contribution (%) of each metal to HI due to the consumption of honey, Figure S1b: Contribution (%) of each metal to HI due to the consumption of bee pollen; Table S1: Description of the codes was reported in Supplementary Materials; Table S2: ICP-MS instrumentation and operating conditions; Table S3: Elemental composition of polyfloral honey found in the present research (Abruzzo) and those found in the literature (μg g^{-1}); Table S4: Elemental composition of polyfloral bee pollen found in the present research (Abruzzo) and those found in the literature (μg g^{-1}).

Author Contributions: Conceptualization, A.C. (Angelo Cichelli), A.C. (Alessandro Chiaudani) and F.F.; methodology, A.C. (Angelo Cichelli), A.C. (Alessandro Chiaudani) and F.F.; formal analysis, A.C. (Alessandro Chiaudani) and F.F.; investigation, A.C. (Ada Consalvo); data curation, A.C. (Alessandro Chiaudani) and F.F.; writing—original draft preparation, A.C. (Alessandro Chiaudani) and F.F.; writing—review and editing, A.C. (Angelo Cichelli), A.C. (Alessandro Chiaudani), F.F. and A.C. (Alessandro Chiaudani); visualization, A.C. (Alessandro Chiaudani) and F.F.; supervision, A.C. (Angelo Cichelli), A.C. (Alessandro Chiaudani) and F.F; project administration, A.C. (Angelo Cichelli); funding acquisition, A.C. (Angelo Cichelli) All authors have read and agreed to the published version of the manuscript.

Funding: This research was conducted within the activities of the RTDA contract cofunded by PON "Ricerca e innovazione" 2014–2020 (PON R&I FSE-REACT EU), Azione IV.6 "Contratti di ricercar su tematiche Green".

Institutional Review Board Statement: Not applicable.

Informed Consent Statement: Not applicable.

Data Availability Statement: The original contributions presented in the study are included in the article/Supplementary Materials; further inquiries can be directed to the corresponding author.

Acknowledgments: The authors are grateful to Polo Agire and Donatantonio De Falcis lamented Inspirer of the Project and Diego Di Curzio for the helpful on the Corine Land Cover maps elaboration.

Conflicts of Interest: The authors declare no conflicts of interest.

References

1. Boukraâ, L. *Honey in Traditional and Modern Medicine*; CRC Press: Boca Raton, FL, USA, 2022.
2. Scivicco, M.; Squillante, J.; Velotto, S.; Esposito, F.; Cirillo, T.; Severino, L. Dietary Exposure to Heavy Metals through Polyfloral Honey from Campania Region (Italy). *J. Food Compos. Anal.* **2022**, *114*, 104748. [CrossRef]
3. Da Silva, P.M.; Gauche, C.; Gonzaga, L.V.; Costa, A.C.O.; Fett, R. Honey: Chemical Composition, Stability and Authenticity. *Food Chem.* **2016**, *196*, 309–323. [CrossRef]
4. Khan, S.U.; Anjum, S.I.; Rahman, K.; Ansari, M.J.; Khan, W.U.; Kamal, S.; Khattak, B.; Muhammad, A.; Khan, H.U. Honey: Single Food Stuff Comprises Many Drugs. *Saudi J. Biol. Sci.* **2018**, *25*, 320–325. [CrossRef]
5. Sakač, M.B.; Jovanov, P.T.; Marić, A.Z.; Pezo, L.L.; Kevrešan, Ž.S.; Novaković, A.R.; Nedeljković, N.M. Physicochemical Properties and Mineral Content of Honey Samples from Vojvodina (Republic of Serbia). *Food Chem.* **2019**, *276*, 15–21. [CrossRef]
6. Wang, H.; Li, L.; Lin, X.; Bai, W.; Xiao, G.; Liu, G. Composition, Functional Properties and Safety of Honey: A Review. *J. Sci. Food Agric.* **2023**, *103*, 6767–6779. [CrossRef]
7. Aga, M.B.; Sharma, V.; Dar, A.H.; Dash, K.K.; Singh, A.; Shams, R.; Khan, S.A. Comprehensive Review on Functional and Nutraceutical Properties of Honey. *eFood* **2023**, *4*, e71. [CrossRef]
8. ISMEA. Mercati Api e Miele—News e Analisi—Tendenze. Available online: https://www.ismeamercati.it/flex/cm/pages/ServeBLOB.php/L/IT/IDPagina/12812 (accessed on 3 May 2024).
9. Barbieri, D.; Gabriele, M.; Summa, M.; Colosimo, R.; Leonardi, D.; Domenici, V.; Pucci, L. Antioxidant, Nutraceutical Properties, and Fluorescence Spectral Profiles of Bee Pollen Samples from Different Botanical Origins. *Antioxidants* **2020**, *9*, 1001. [CrossRef]
10. Khalifa, S.A.M.; Elashal, M.H.; Yosri, N.; Du, M.; Musharraf, S.G.; Nahar, L.; Sarker, S.D.; Guo, Z.; Cao, W.; Zou, X.; et al. Bee Pollen: Current Status and Therapeutic Potential. *Nutrients* **2021**, *13*, 1876. [CrossRef]
11. Kostić, A.; Milinčić, D.D.; Barać, M.B.; Shariati, M.A.; Tešić, Ž.L.; Pešić, M.B. The Application of Pollen as a Functional Food and Feed Ingredient—The Present and Perspectives. *Biomolecules* **2020**, *10*, 84. [CrossRef]
12. Mohdaly, A.A.A.; Mahmoud, A.A.; Roby, M.H.H.; Smetanska, I.; Ramadan, M.F. Phenolic Extract from Propolis and Bee Pollen: Composition, Antioxidant and Antibacterial Activities. *J. Food Biochem.* **2015**, *39*, 538–547. [CrossRef]
13. Végh, R.; Csóka, M.; Sörös, C.; Sipos, L. Food Safety Hazards of Bee Pollen—A Review. *Trends Food Sci. Technol.* **2021**, *114*, 490–509. [CrossRef]

14. Atamaleki, A.; Yazdanbakhsh, A.; Fakhri, Y.; Mahdipour, F.; Khodakarim, S.; Mousavi Khaneghah, A. The Concentration of Potentially Toxic Elements (PTEs) in the Onion and Tomato Irrigated by Wastewater: A Systematic Review; Meta-Analysis and Health Risk Assessment. *Food Res. Int.* **2019**, *125*, 108518. [CrossRef]
15. World Health Organization; Food and Agriculture Organization of the United Nations; International Atomic Energy Agency. *Trace Elements in Human Nutrition and Health*; World Health Organization: Geneva, Switzerland, 1996; p. 343.
16. Squadrone, S.; Brizio, P.; Stella, C.; Pederiva, S.; Brusa, F.; Mogliotti, P.; Garrone, A.; Abete, M.C. Trace and Rare Earth Elements in Monofloral and Multifloral Honeys from Northwestern Italy; A First Attempt of Characterization by a Multi-Elemental Profile. *J. Trace Elem. Med. Biol.* **2020**, *61*, 126556. [CrossRef]
17. Walker, C.H. Neurotoxic Pesticides and Behavioural Effects Upon Birds. *Ecotoxicology* **2003**, *12*, 307–316. [CrossRef]
18. Perna, A.M.; Grassi, G.; Gambacorta, E.; Simonetti, A. Minerals Content in Basilicata Region (Southern Italy) Honeys from Areas with Different Anthropic Impact. *Int. J. Food Sci. Technol.* **2021**, *56*, 4465–4472. [CrossRef]
19. Villanueva, R.; Bustamante, P. Composition in Essential and Non-Essential Elements of Early Stages of Cephalopods and Dietary Effects on the Elemental Profiles of Octopus Vulgaris Paralarvae. *Aquaculture* **2006**, *261*, 225–240. [CrossRef]
20. Alqarni, A.S.; Owayss, A.A.; Mahmoud, A.A.; Hannan, M.A. Mineral Content and Physical Properties of Local and Imported Honeys in Saudi Arabia. *J. Saudi Chem. Soc.* **2014**, *18*, 618–625. [CrossRef]
21. Tchounwou, P.B.; Yedjou, C.G.; Patlolla, A.K.; Sutton, D.J. Heavy Metal Toxicity and the Environment. In *Molecular, Clinical and Environmental Toxicology. Experientia Supplementum*; Springer: Basel, Switzerland, 2012; Volume 101, pp. 133–164.
22. Briffa, J.; Sinagra, E.; Blundell, R. Heavy Metal Pollution in the Environment and Their Toxicological Effects on Humans. *Heliyon* **2020**, *6*, e04691. [CrossRef]
23. Scutarașu, E.C.; Trincă, L.C. Heavy Metals in Foods and Beverages: Global Situation, Health Risks and Reduction Methods. *Foods* **2023**, *12*, 3340. [CrossRef]
24. Gautam, P.K.; Gautam, R.K.; Chattopadhyaya, M.C.; Banerjee, S.; Chattopadhyaya, M.C.; Pandey, J.D. *Heavy Metals in the Environment: Fate, Transport, Toxicity and Remediation Technologies*; Nova Science Publishers: Hauppauge, NY, USA, 2016.
25. Squadrone, S.; Brizio, P.; Stella, C.; Mantia, M.; Pederiva, S.; Brusa, F.; Mogliotti, P.; Garrone, A.; Abete, M.C. Trace Elements and Rare Earth Elements in Honeys from the Balkans, Kazakhstan, Italy, South America, and Tanzania. *Environ. Sci. Pollut. Res.* **2020**, *27*, 12646–12657. [CrossRef]
26. Zambelli, B.; Uversky, V.N.; Ciurli, S. Nickel Impact on Human Health: An Intrinsic Disorder Perspective. *Biochim. Biophys. Acta Proteins Proteom.* **2016**, *1864*, 1714–1731. [CrossRef]
27. Erdoğan, A.; Şeker, M.E.; Kahraman, S.D. Evaluation of Environmental and Nutritional Aspects of Bee Pollen Samples Collected from East Black Sea Region, Turkey, via Elemental Analysis by ICP-MS. *Biol. Trace Elem. Res.* **2023**, *201*, 1488–1502. [CrossRef]
28. Sharma, A.; Pant, K.; Brar, D.S.; Thakur, A.; Nanda, V. A Review on Api-Products: Current Scenario of Potential Contaminants and Their Food Safety Concerns. *Food Control* **2023**, *145*, 109499. [CrossRef]
29. World Health Organization. Lead Poisoning. Available online: https://www.who.int/news-room/fact-sheets/detail/lead-poisoning-and-health (accessed on 3 May 2024).
30. European Commission. *EU Regulation 2023/915*; European Commission: Brussels, Belgium, 2023.
31. Liolios, V.; Tananaki, C.; Papaioannou, A.; Kanelis, D.; Rodopoulou, M.A.; Argena, N. Mineral Content in Monofloral Bee Pollen: Investigation of the Effect of the Botanical and Geographical Origin. *J. Food Meas. Charact.* **2019**, *13*, 1674–1682. [CrossRef]
32. Morgano, M.A.; Martins, M.C.T.; Rabonato, L.C.; Milani, R.F.; Yotsuyanagi, K.; Rodriguez-Amaya, D.B. A Comprehensive Investigation of the Mineral Composition of Brazilian Bee Pollen: Geographic and Seasonal Variations and Contribution to Human Diet. *Artic. J. Braz. Chem. Soc.* **2012**, *23*, 727–736. [CrossRef]
33. Wang, H.; Inukai, Y.; Yamauchi, A. Root Development and Nutrient Uptake. *CRC Crit. Rev. Plant Sci.* **2006**, *25*, 279–301. [CrossRef]
34. Porrini, C.; Medrzycki, P. Avvelenamenti Da Pesticidi. In *Patologia e Avversità Dell'alveare*; Springer: Milan, Italy, 2014; pp. 293–323.
35. Perugini, M.; Manera, M.; Grotta, L.; Abete, M.C.; Tarasco, R.; Amorena, M. Heavy Metal (Hg, Cr, Cd, and Pb) Contamination in Urban Areas and Wildlife Reserves: Honeybees as Bioindicators. *Biol. Trace Elem. Res.* **2011**, *140*, 170–176. [CrossRef]
36. Couvillon, M.J.; Ratnieks, F.L.W. Environmental Consultancy: Dancing Bee Bioindicators to Evaluate Landscape "Health". *Front. Ecol. Evol.* **2015**, *3*, 44. [CrossRef]
37. Nowak, A.; Nowak, I. Review of Harmful Chemical Pollutants of Environmental Origin in Honey and Bee Products. *Crit. Rev. Food Sci. Nutr.* **2023**, *63*, 5094–5116. [CrossRef]
38. González-Alcaraz, M.N.; Malheiro, C.; Cardoso, D.N.; Prodana, M.; Morgado, R.G.; van Gestel, C.A.M.; Loureiro, S. Bioaccumulation and Toxicity of Organic Chemicals in Terrestrial Invertebrates. In *Handbook of Environmental Chemistry*; Springer Science and Business Media Deutschland GmbH: Berlin/Heidelberg, Germany, 2020; Volume 100, pp. 149–189.
39. Borsuk, G.; Sulborska, A.; Stawiarz, E.; Olszewski, K.; Wiącek, D.; Ramzi, N.; Nawrocka, A.; Jędryczka, M. Capacity of Honeybees to Remove Heavy Metals from Nectar and Excrete the Contaminants from Their Bodies. *Apidologie* **2021**, *52*, 1098–1111. [CrossRef]
40. Conti, M.E.; Canepari, S.; Finoia, M.G.; Mele, G.; Astolfi, M.L. Characterization of Italian Multifloral Honeys on the Basis of Their Mineral Content and Some Typical Quality Parameters. *J. Food Compos. Anal.* **2018**, *74*, 102–113. [CrossRef]
41. Dżugan, M.; Wesołowska, M.; Zaguła, G.; Kaczmarski, M.; Czernicka, M.; Puchalski, C. Honeybees (*Apis mellifera*) as a Biological Barrier for Contamination of Honey by Environmental Toxic Metals. *Environ. Monit. Assess.* **2018**, *190*, 101. [CrossRef]
42. Ruschioni, S.; Riolo, P.; Minuz, R.L.; Stefano, M.; Cannella, M.; Porrini, C.; Isidoro, N. Biomonitoring with Honeybees of Heavy Metals and Pesticides in Nature Reserves of the Marche Region (Italy). *Biol. Trace Elem. Res.* **2013**, *154*, 226–233. [CrossRef]

43. Formicki, G.; Greń, A.; Stawarz, R.; Zyśk, B.; Gał, A. Metal Content in Honey, Propolis, Wax, and Bee Pollen and Implications for Metal Monitoring. *Pol. J. Environ. Stud.* **2013**, *22*, 99–106.
44. Salkova, D.; Panayotova-Pencheva, M. Honey Bees and Their Products as Indicators of Environmental Pollution: A Review. *Agric. Sci. Technol.* **2016**, *8*, 175–182. [CrossRef]
45. Copernicus Land Monitoring Service Data Viewer. Available online: https://land.copernicus.eu/en/map-viewer (accessed on 8 May 2024).
46. Batista, B.L.; da Silva, L.R.S.; Rocha, B.A.; Rodrigues, J.L.; Berretta-Silva, A.A.; Bonates, T.O.; Gomes, V.S.D.; Barbosa, R.M.; Barbosa, F. Multi-Element Determination in Brazilian Honey Samples by Inductively Coupled Plasma Mass Spectrometry and Estimation of Geographic Origin with Data Mining Techniques. *Food Res. Int.* **2012**, *49*, 209–215. [CrossRef]
47. Kalaycıoğlu, Z.; Kaygusuz, H.; Döker, S.; Kolaylı, S.; Erim, F.B. Characterization of Turkish Honeybee Pollens by Principal Component Analysis Based on Their Individual Organic Acids, Sugars, Minerals, and Antioxidant Activities. *LWT* **2017**, *84*, 402–408. [CrossRef]
48. Conti, M.E.; Astolfi, M.L.; Finoia, M.G.; Massimi, L.; Canepari, S. Biomonitoring of Element Contamination in Bees and Beehive Products in the Rome Province (Italy). *Environ. Sci. Pollut. Res.* **2022**, *29*, 36057–36074. [CrossRef]
49. Hague, T.; Petroczi, A.; Andrews, P.L.R.; Barker, J.; Naughton, D.P. Determination of Metal Ion Content of Beverages and Estimation of Target Hazard Quotients: A Comparative Study. *Chem. Cent. J.* **2008**, *2*, 13. [CrossRef]
50. Ozyigit, I.I.; Karahan, F.; Yalcin, I.E.; Hocaoglu-Ozyigit, A.; Ilcim, A. Heavy Metals and Trace Elements Detected in the Leaves of Medicinal Plants Collected in the Southeast Part of Turkey. *Arab. J. Geosci.* **2021**, *15*, 27. [CrossRef]
51. US EPA. Regional Screening Levels (RSLs)—User's Guide. Available online: https://www.epa.gov/risk/regional-screening-levels-rsls-users-guide (accessed on 3 May 2024).
52. Çobanoğlu, D.N.; Kizilpinar Temizer, İ.; Candan, E.D.; Yolcu, U.; Güder, A. Evaluation of the Nutritional Value of Bee Pollen by Palynological, Antioxidant, Antimicrobial, and Elemental Characteristics. *Eur. Food Res. Technol.* **2023**, *249*, 307–325. [CrossRef]
53. Winiarska-Mieczan, A.; Wargocka, B.; Jachimowicz, K.; Baranowska-Wójcik, E.; Kwiatkowska, K.; Kwiecień, M. Evaluation of Consumer Safety of Polish Honey—The Content of Cd and Pb in Multifloral, Monofloral and Honeydew Honeys. *Biol. Trace Elem. Res.* **2021**, *199*, 4370–4383. [CrossRef]
54. Leclercq, C.; Arcella, D.; Piccinelli, R.; Sette, S.; Le Donne, C. The Italian National Food Consumption Survey INRAN-SCAI 2005-06: Main Results: In Terms of Food Consumption. *Public Health Nutr.* **2009**, *12*, 2504–2532. [CrossRef]
55. Fakhri, Y.; Abtahi, M.; Atamaleki, A.; Raoofi, A.; Atabati, H.; Asadi, A.; Miri, A.; Shamloo, E.; Alinejad, A.; Keramati, H.; et al. The Concentration of Potentially Toxic Elements (PTEs) in Honey: A Global Systematic Review and Meta-Analysis and Risk Assessment. *Trends Food Sci. Technol.* **2019**, *91*, 498–506. [CrossRef]
56. Zafeiraki, E.; Kasiotis, K.M.; Nisianakis, P.; Manea-Karga, E.; Machera, K. Occurrence and Human Health Risk Assessment of Mineral Elements and Pesticides Residues in Bee Pollen. *Food Chem. Toxicol.* **2022**, *161*, 112826. [CrossRef]
57. Komosinska-Vassev, K.; Olczyk, P.; Kaźmierczak, J.; Mencner, L.; Olczyk, K. Bee Pollen: Chemical Composition and Therapeutic Application. *Evid.-Based Complement. Altern. Med.* **2015**, *2015*, 297425. [CrossRef]
58. Pohl, P.; Dzimitrowicz, A.; Greda, K.; Jamroz, P.; Lesniewicz, A.; Szymczycha-Madeja, A.; Welna, M. Element Analysis of Bee-Collected Pollen and Bee Bread by Atomic and Mass Spectrometry—Methodological Development in Addition to Environmental and Nutritional Aspects. *TrAC—Trends Anal. Chem.* **2020**, *128*, 115922. [CrossRef]
59. Sevin, S.; Tutun, H.; Yipel, M.; Aluç, Y.; Ekici, H. Concentration of Essential and Non-Essential Elements and Carcinogenic/Non-Carcinogenic Health Risk Assessment of Commercial Bee Pollens from Turkey. *J. Trace Elem. Med. Biol.* **2023**, *75*, 127104. [CrossRef]
60. Bat, L.; Şahin, F.; Öztekin, A.; Arici, E.; Yardim, Ö. Assessment of Cd, Hg, Pb, Cu and Zn Amounts in Muscles of Cyprinus Carpio from Karasu Stream, Sinop. *Curr. Agric. Res. J.* **2019**, *7*, 171–180. [CrossRef]
61. Obiora, S.C.; Chukwu, A.; Chibuike, G.; Nwegbu, A.N. Potentially Harmful Elements and Their Health Implications in Cultivable Soils and Food Crops around Lead-Zinc Mines in Ishiagu, Southeastern Nigeria. *J. Geochem. Explor.* **2019**, *204*, 289–296. [CrossRef]
62. Real, M.I.H.; Azam, H.M.; Majed, N. Consumption of Heavy Metal Contaminated Foods and Associated Risks in Bangladesh. *Environ. Monit. Assess.* **2017**, *189*, 651. [CrossRef]
63. Ullah, R.; Jan, F.A.; Gulab, H.; Saleem, S.; Ullah, N.; Wajidullah. Metals Contents in Honey, Beeswax and Bees and Human Health Risk Assessment Due to Consumption of Honey: A Case Study from Selected Districts in Khyber Pakhtunkhwa, Pakistan. *Arch. Environ. Contam. Toxicol.* **2022**, *82*, 341–354. [CrossRef]
64. Gebeyehu, H.R.; Bayissa, L.D. Levels of Heavy Metals in Soil and Vegetables and Associated Health Risks in Mojo Area, Ethiopia. *PLoS ONE* **2020**, *15*, e0227883. [CrossRef]
65. Nduka, J.K.; Kelle, H.I.; Amuka, J.O. Health Risk Assessment of Cadmium, Chromium and Nickel from Car Paint Dust from Used Automobiles at Auto-Panel Workshops in Nigeria. *Toxicol. Rep.* **2019**, *6*, 449–456. [CrossRef]
66. U.S. Department of Energy (DOE). The Risk Assessment Information System. Available online: https://rais.ornl.gov/cgi-bin/tools/TOX_search?select=chemtox (accessed on 3 May 2024).
67. De Souza, I.D.; Melo, E.S.P.; Nascimento, V.A.; Pereira, H.S.; Silva, K.R.N.; Espindola, P.R.; Tschinkel, P.F.S.; Ramos, E.M.; Reis, F.J.M.; Ramos, I.B.; et al. Potential Health Risks of Macro- And Microelements in Commercial Medicinal Plants Used to Treatment of Diabetes. *Biomed. Res. Int.* **2021**, *2021*, 6678931. [CrossRef]

68. US EPA; Superfund Health Risk Technical Support Center. *Provisional Peer Reviewed Toxicity Values for Iron and Compounds (CASRN 7439-89-6)*; US EPA: Washington, DC, USA, 2006.

69. US EPA; Superfund Health Risk Technical Support Center. *Provisional Peer-Reviewed Toxicity Values for Rubidium Compounds (CASRN 7440-17-7, Rubidium) (CASRN 7791-11-9, Rubidium Chloride) (CASRN 1310-82-3, Rubidium Hydroxide) (CASRN 7790-29-6, Rubidium Iodide)*; US EPA: Washington, DC, USA, 2006.

70. Keith, S.; Faroon, O.; Roney, N.; Scinicariello, F.; Wilbur, S.; Ingerman, L.; Llados, F.; Plewak, D.; Wohlers, D.; Diamond, G. *Toxicological Profile for Uranium*; U.S. Department of Health & Human Services: Washington, DC, USA, 2013.

71. Health Canada. *Federal Contaminated Site Risk Assessment in Canada, Part V: Guidance on Complex Human Health Detailed Quantitative Risk Assessment for Chemicals (DQRACHEM)*; Health Canada: Ottawa, ON, USA, 2010.

72. Metsalu, T.; Vilo, J. ClustVis: A Web Tool for Visualizing Clustering of Multivariate Data Using Principal Component Analysis and Heatmap. *Nucleic Acids Res.* **2015**, *43*, W566–W570. [CrossRef]

73. Astolfi, V.; Kaškonienė, V.; Kaškonas, P.; Barčauskaitė, K.; Maruška, A. Comparison of Physicochemical Properties of Bee Pollen with Other Bee Products. *Biomolecules* **2019**, *9*, 819. [CrossRef]

74. Astolfi, M.L.; Conti, M.E.; Messi, M.; Marconi, E. Probiotics as a Promising Prophylactic Tool to Reduce Levels of Toxic or Potentially Toxic Elements in Bees. *Chemosphere* **2022**, *308*, 136261. [CrossRef]

75. Quinto, M.; Miedico, O.; Spadaccino, G.; Paglia, G.; Mangiacotti, M.; Li, D.; Centonze, D.; Chiaravalle, A.E. Characterization, Chemometric Evaluation, and Human Health-Related Aspects of Essential and Toxic Elements in Italian Honey Samples by Inductively Coupled Plasma Mass Spectrometry. *Environ. Sci. Pollut. Res.* **2016**, *23*, 25374–25384. [CrossRef]

76. Di Bella, G.; Lo Turco, V.; Potortì, A.G.; Bua, G.D.; Fede, M.R.; Dugo, G. Geographical Discrimination of Italian Honey by Multi-Element Analysis with a Chemometric Approach. *J. Food Compos. Anal.* **2015**, *44*, 25–35. [CrossRef]

77. Ru, Q.M.; Feng, Q.; He, J.Z. Risk Assessment of Heavy Metals in Honey Consumed in Zhejiang Province, Southeastern China. *Food Chem. Toxicol.* **2013**, *53*, 256–262. [CrossRef]

78. Perna, A.; Intaglietta, I.; Simonetti, A.; Gambacorta, E. Metals in Honeys from Different Areas of Southern Italy. *Bull. Environ. Contam. Toxicol.* **2014**, *92*, 253–258. [CrossRef]

79. Ligor, M.; Kowalkowski, T.; Buszewski, B. Comparative Study of the Potentially Toxic Elements and Essential Microelements in Honey Depending on the Geographic Origin. *Molecules* **2022**, *27*, 5474. [CrossRef]

80. Squadrone, S.; Brizio, P.; Stella, C.; Mantia, M.; Battuello, M.; Nurra, N.; Sartor, R.M.; Orusa, R.; Robetto, S.; Brusa, F.; et al. Rare Earth Elements in Marine and Terrestrial Matrices of Northwestern Italy: Implications for Food Safety and Human Health. *Sci. Total Environ.* **2019**, *660*, 1383–1391. [CrossRef]

81. Pisani, A.; Protano, G.; Riccobono, F. Minor and Trace Elements in Different Honey Types Produced in Siena County (Italy). *Food Chem.* **2008**, *107*, 1553–1560. [CrossRef]

82. Meli, M.A.; Desideri, D.; Roselli, C.; Benedetti, C.; Feduzi, L. Essential and Toxic Elements in Honeys from a Region of Central Italy. *J. Toxicol. Environ. Health—Part. A Curr. Issues* **2015**, *78*, 617–627. [CrossRef]

83. Bogdanov, S. The pollen Book. In *Pollen: Collection, Harvest, Composition, Quality*; Chapter 1; Bogdanov, S., Ed.; Bee Product Science: Muehlethurnen, Switzerland, 2017.

84. Conti, M.E.; Botrè, F. Honeybees and Their Products as Potential Bioindicators of Heavy Metals Contamination. *Environ. Monit. Assess.* **2001**, *69*, 267–282. [CrossRef]

85. Lambert, O.; Piroux, M.; Puyo, S.; Thorin, C.; Larhantec, M.; Delbac, F.; Pouliquen, H. Bees, Honey and Pollen as Sentinels for Lead Environmental Contamination. *Environ. Pollut.* **2012**, *170*, 254–259. [CrossRef]

86. Roman, A. Concentration of Chosen Trace Elements of Toxic Properties in Bee Pollen Loads. *Pol. J. Environ. Stud.* **2009**, *18*, 265–272.

87. Begum, W.; Rai, S.; Banerjee, S.; Bhattacharjee, S.; Mondal, M.H.; Bhattarai, A.; Saha, B. A Comprehensive Review on the Sources, Essentiality and Toxicological Profile of Nickel. *RSC Adv.* **2022**, *12*, 9139–9153. [CrossRef]

88. Pietrelli, L.; Menegoni, P.; Papetti, P. Bioaccumulation of Heavy Metals by Herbaceous Species Grown in Urban and Rural Sites. *Water Air Soil. Pollut.* **2022**, *233*, 141. [CrossRef]

89. Catalano, P.; Della Sala, F.; Cavaliere, M.; Caputo, C.; Pecoraro, D.; Crispino, G.; Lettera, S.; Caioni, G.; Esposito, M.; Verre, A.; et al. Use of Honey Bees and Hive Products as Bioindicators to Assess Environmental Contamination in Targeted Areas of the Campania Region (Italy). *Animals* **2024**, *14*, 1446. [CrossRef]

90. Farias, R.A.; Nunes, C.N.; Quináia, S.P. Bees Reflect Better on Their Ecosystem Health than Their Products. *Environ. Sci. Pollut. Res.* **2023**, *30*, 79617–79626. [CrossRef]

91. Smith, K.E.; Weis, D. Evaluating Spatiotemporal Resolution of Trace Element Concentrations and Pb Isotopic Compositions of Honeybees and Hive Products as Biomonitors for Urban Metal Distribution. *GeoHealth* **2020**, *4*, e2020GH000264. [CrossRef]

92. Wong, C.; Roberts, S.M.; Saab, I.N. Review of Regulatory Reference Values and Background Levels for Heavy Metals in the Human Diet. *Regul. Toxicol. Pharmacol.* **2022**, *130*, 105122. [CrossRef]

93. Beshaw, T.; Demssie, K.; Leka, I. Levels and Health Risk Assessment of Trace Metals in Honey from Different Districts of Bench Sheko Zone, Southwest Ethiopia. *Heliyon* **2022**, *8*, e10535. [CrossRef]

94. Obasi, N.A.; Aloke, C.; Obasi, S.E.; Famurewa, A.C.; Ogbu, P.N.; Onyeji, G.N. Elemental Composition and Associated Health Risk of Honey Obtained from Apiary Farms in Southeast Nigeria. *J. Food Prot.* **2020**, *83*, 1745–1756. [CrossRef]

95. Orisakwe, O.E.; Ozoani, H.A.; Nwaogazie, I.L.; Ezejiofor, A.N. Probabilistic Health Risk Assessment of Heavy Metals in Honey, Manihot Esculenta, and Vernonia Amygdalina Consumed in Enugu State, Nigeria. *Environ. Monit. Assess.* **2019**, *191*, 424. [CrossRef]

96. Pipoyan, D.; Stepanyan, S.; Beglaryan, M.; Stepanyan, S.; Asmaryan, S.; Hovsepyan, A.; Merendino, N. Carcinogenic and Non-Carcinogenic Risk Assessment of Trace Elements and POPs in Honey from Shirak and Syunik Regions of Armenia. *Chemosphere* **2020**, *239*, 124809. [CrossRef]

97. AL-Alam, J.; Chbani, A.; Faljoun, Z.; Millet, M. The Use of Vegetation, Bees, and Snails as Important Tools for the Biomonitoring of Atmospheric Pollution—A Review. *Environ. Sci. Pollut. Res.* **2019**, *26*, 9391–9408. [CrossRef]

Article

Flavor Chemical Research on Different Bee Pollen Varieties Using Fast E-Nose and E-Tongue Technology

Chenshuo Liu [1,2], Enning Zhou [3], Yuying Zhu [3], Qiangqiang Li [3,*] and Liming Wu [1,*]

1   Hainan Academy of Agricultural Sciences, Haikou 571100, China
2   School of Food and Health, Beijing Technology and Business University, Beijing 100048, China
3   Institute of Apicultural Research, Chinese Academy of Agricultural Sciences, Beijing 100093, China
*   Correspondence: qiangqiangli1991@163.com (Q.L.); apiswu@126.com (L.W.)

Abstract: Bee pollen, derived from various plant sources, is renowned for its nutritional and bioactive properties, aroma, and taste. This study examined the bee pollen with the highest yield in China obtained from four plant species, namely *Brassica campestris* (Bc), *Nelumbo nucifera* (Nn), *Camellia japonica* (Cj), and *Fagopyrum esculentum* (Fe), using fast e-nose and e-tongue technology to analyze their flavor chemistry. Results showed substantial differences in scent profiles among the varieties, with distinct odor compounds identified for each, including n-butanol, decanal, and ethanol, in Bc, Nn, and Cj, respectively. The primary odorants in Fe consist of E-2-hexen-1-ol and (Z)-3-hexen-1-ol. Additionally, e-tongue analysis revealed seven distinct tastes in bee pollen samples: AHS, PKS, CTS, NMS, CPS, ANS, and SCS, with variations in intensity across each taste. The study also found correlations between taste components and specific odor compounds, providing insights for enhancing product quality control in bee pollen processing.

Keywords: bee pollen; e-nose; e-tongue; flavor chemistry

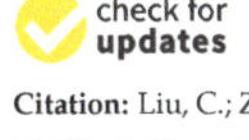

Citation: Liu, C.; Zhou, E.; Zhu, Y.; Li, Q.; Wu, L. Flavor Chemical Research on Different Bee Pollen Varieties Using Fast E-Nose and E-Tongue Technology. *Foods* **2024**, *13*, 1022. https://doi.org/10.3390/foods13071022

Academic Editor: Olga Escuredo

Received: 6 March 2024
Revised: 21 March 2024
Accepted: 23 March 2024
Published: 27 March 2024

1. Introduction

Bee pollen is formed via pollen collection by bees, which is then combined with saliva secretions and nectar. Due to its diverse health benefits, such as disease prevention, bee pollen has received heightened attention in food processing fields [1]. Bee pollen is rich in carbohydrates, essential amino acids, unsaturated fatty acids, vitamins, and various micronutrients [2]. It is a 'complete food' containing all amino acids necessary in the human diet [3]. Moreover, it demonstrates significant biological activities, including antioxidant, anti-inflammatory, and hypolipidemic effects [4,5]. Recent studies have also demonstrated the regulatory potential of bee pollen polysaccharides on intestinal microbiota [6]. In addition, recent research suggests that bee pollen may be used as a biofunctional ingredient for enhancing product quality, potentially incorporated into yogurt, cheese, bread, and fermented beverages. Bee pollen can be utilized as an additive in biomedical formulations for bioprinting, biopolymers, tissue engineering, and nanoparticle formulation [7]. Furthermore, the extensive number of plant origins and the diverse array of resulting flavors make it critical to conduct flavor chemistry investigations on bee pollen, underscoring its profound significance.

Electronic nose (e-nose) and electronic tongue (e-tongue) technologies are two essential branches of contemporary sensor technology, mimicking the olfactory and gustatory systems of humans to differentiate chemical constituents within samples. Using the overall characteristic response signal of a sample to simulate identification and conduct quantitative and qualitative analysis has been widely utilized across various domains due to its speed, ease of operation, and reproducibility. The utilization of e-tongue and e-nose in food quality control and production monitoring is prevalent. The applications of the e-tongue and e-nose, with prediction accuracies ranging from 80% to 96%, were significant in the field of food analysis [8]. The integration of diverse intelligent sensory algorithms has become ubiquitous, with particular emphasis placed on the incorporation of e-tongue and e-nose technologies. The HERACLES e-nose instrument utilizes rapid gas chromatography technology, significantly enhancing the efficiency of qualitative and quantitative analysis for complex odor samples. The ASTREE e-tongue system relies on the measurement of potential differences across sensors directly contacting liquids, allowing the assessment of taste variations among products and formulations. According to Xia et al., an e-nose can effectively collect information on aroma compounds using a sensor array, allowing the identification of changes in tea aroma during processing and determining the quality of tea [9]. Based on research by Estivi et al., the alkaloid content of lupin seeds debittered using different solvents and ultrasound for varying soaking times was determined, while the taste profile was assessed using e-tongue technology [10]. Banerjee et al. applied e-nose and e-tongue systems to evaluate black tea quality and determined that the integrated systems achieved higher classification accuracy relative to individual systems [11].

The utilization of e-nose and e-tongue technologies is immensely important for assessing bee pollen quality and flavor, due to their rapid and non-invasive characteristics. E-noses often consist of gas-sensitive sensors that selectively respond to volatile compounds found in different chemicals. Upon contact with these sensors, the conductivity of the volatile compounds from bee pollen undergoes changes, which identifies and differentiates odor components via signal processing and pattern recognition techniques. The e-tongue can be utilized to assess taste characteristics in bee pollen samples, potentially influencing the perception of their aroma. Consequently, the data obtained from the e-tongue can offer additional information to facilitate a more comprehensive understanding of the sensory properties of these samples.

Therefore, this study selected the bee pollen with the highest yield in China obtained from four plant species, including *Brassica campestris* (Bc), *Nelumbo nucifera* (Nn), *Camellia japonica* (Cj), and *Fagopyrum esculentum* (Fe), and conducted a flavor chemical research based on fast e-nose and e-tongue technology. Using e-nose and e-tongue systems for qualitative and quantitative analysis, we aimed to discern variations in odor profiles and taste profiles, respectively, across the varieties of bee pollen. This study aims to establish a solid scientific foundation for the processing and use of bee pollen while emphasizing the importance of quality control measures for its resulting products.

2. Materials and Methods

2.1. Sample Collection and Reagent Acquisition

The bee pollen samples were collected during respective flowering season of four plant species: *Brassica campestris* (Bc), *Nelumbo nucifera* (Nn), *Camellia japonica* (Cj), and *Fagopyrum esculentum* (Fe), cultivated at the apiary of the Institute of Apicultural Research of Chinese Academy of Agricultural Sciences (IAR, CAAS, Beijing, China). The collected samples underwent grinding and freeze-drying to obtain a powder, stored at -20 °C. Various n-alkane (nC6 to nC16) standards were purchased from ZZBIO Co., Ltd. (Shanghai, China) for GC analysis.

2.2. Scanning Electron Microscopy Examination on Bee Pollen Samples

Prior to performing scanning electron microscopy (SEM) analysis, bee pollen powders were dispersed in water and spread evenly onto tin foil. The samples underwent a drying process before mounting on metal stubs. A thin layer of gold was coated onto the samples before observation using a Hitachi S-750 SEM system manufactured by the Hitachi Company, Tokyo, Japan.

2.3. Preparation of Bee Pollen Samples for Analysis

A sample of 0.2 g of bee pollen powder was weighed into a headspace vial with a capacity of 20 mL, designed for e-noses. The vial was sealed using a PTFE liner and prepared as five parallel samples. Subsequently, the prepared samples were placed onto an automated sampler device to perform analysis utilizing the e-nose.

A sample of 5 g of bee pollen powder was dissolved in 100 mL of 40% ethanol via ultrasonic treatment. Subsequently, the prepared pollen solution was filtered using filter paper and carefully transferred into a specialized e-tongue beaker with a capacity of 25 mL. The e-tongue was allowed to measure the solution accurately.

2.4. E-Nose Analysis

Samples were analyzed using the HERACLES NEO ultra-rapid gas chromatography e-nose, following the experimental conditions outlined in Table 1. Data processing was performed using AlphaSoft 2023 software. Calibration was conducted employing a standard solution of n-alkanes (nC6 to nC16), and the retention times were converted to retention indices for qualitative analysis of compounds referring to the AroChemBase database.

Table 1. Heracles NEO instrument parameter configurations.

Parameters	Value
Headspace vial	20 mL
Sample amount	0.2 g
Incubation temperature	80 °C
Incubation time	20 min
Inlet volume	5000 µL
Inlet speed	125 µL/s
Inlet temperature	200 °C
Inlet duration	45 s
Initial trap temperature	40 °C
Split mode	10 mL/min
Injection duration	50 s
Final trap temperature	240 °C
Initial column temperature	40 °C (30 s)
RAMP	0.5 °C/s −60 °C (0 s) 2.0 °C/s −250 °C (15 s)
Acquisition time	180 s
Detector temperature	260 °C
FID	12

2.5. E-Tongue Analysis

An e-tongue was used to identify taste indicators across diverse bee pollen samples. Prior to sample measurement, sensor activation, calibration, and diagnosis were performed to guarantee a consistent sensor status. The e-tongue system incorporated the 6th-generation sensor system, consisting of AHS, ANS, SCS, CTS, NMS, PKS, and CPS sensors alongside a standard reference electrode (Ag/AgCl), totaling seven sensors. Among them, AHS, ANS, SCS, CTS, and NMS exhibited sensitivity towards taste attributes of sourness, sweetness, bitterness, saltiness, and umami, respectively. PKS and CPS functioned as composite sensors [12]. To ensure precise detection, the experimental sample volume was 25 mL, while the sampling time was set at 120 s. It was observed that performing three repeated measurements yielded optimal testing conditions for data analysis.

2.6. Statistical Analysis

The statistical methodology employed in this study was consistent with the approach described in our previous publication [13]. A *t*-test was performed using SPSS version 21.0 (IBM Co., Armonk, NY, USA). The dataset utilized in the *t*-test followed a Gaussian distribution and exhibited homogeneity of variance, guaranteeing the validity of the test results. A significance level of 0.05 or lower was considered statistically significant, indicating a notable distinction between the two designated groups. For principal component analysis

(PCA) and orthogonal partial least-squares discriminant analysis (OPLS-DA), we utilized SIMCA-P version 13.0 software (SSB Co., Svedala, Sweden).

3. Results and Discussion

3.1. Bee Pollen Morphology Analysis

A commonly utilized approach for determining the botanical origin of pollen loads is microscopic pollen analysis, as the size, shape, and surface properties of pollen grains are specific to particular plant species [14]. The micro-morphology of bee pollen is essential for its contribution to plant reproduction, and is linked to the pollination mechanism, genetic diversity, and adaptability of specific plants. The morphology, size, exine ornamentation type, and germination pore type are primary indicators for pollen examination using electron microscopy. As documented, pollen grains have intricate patterns along their outer walls, with the pollen coat seamlessly enveloping the outer wall layer and intricately sculpted surfaces [15,16]. Figure 1 illustrates electron microscopic images showcasing four distinct varieties of bee pollen. Upon examination of the electron microscope images, Bc exhibited pollen grains to be as follows: monad, radial, isopolar, tricolporate; reticulate, homobrochate, brochi coarse, lumina ca. 1.0 μm wide, muri very thin, simplicolumellate, columellae baculae shaped; colpus as long as grain, wide; polar shape circular; grains prolate to subprolate, ca. 30.0 μm long × 24.0 μm wide. The Nn sample exhibited pollen grains to be as follows: monad, radial, isopolar, tricolpate; verrucate, verrucae fine resembling small baculae; colpus ¾ as long as grain, thin; polar shape circular; grains subprolate, ca. 68.0 μm long × 60.0 μm wide. The Cj sample exhibited pollen grains to be as follows: monad, radial, isopolar, tricolporate; reticulate, brochi fine, muri simplicolumellate, baculae shaped; colpus as long as grain; pore inconspicuous, slightly protruding; polar shape triangular; grains suboblate, ca. 35.0 μm long × 36.0 μm wide. The Fe sample exhibited pollen grains to be as follows: monad, radial, isopolar, tricolporate; baculate, baculae coarse, lumina ca. 1.0 to 2.0 μm wide; colpus as long as grain, thin, ends acute; pore lalongate, depressed; grains subprolate, ca. 68.0 μm long × 36.0 μm wide. When examining the four varieties of bee pollen, there was a subtle disparity in hue, while most were predominantly yellow. The coloration of the Bc sample had enhanced vibrancy and luminosity, while the Fe sample's color appeared comparatively deeper.

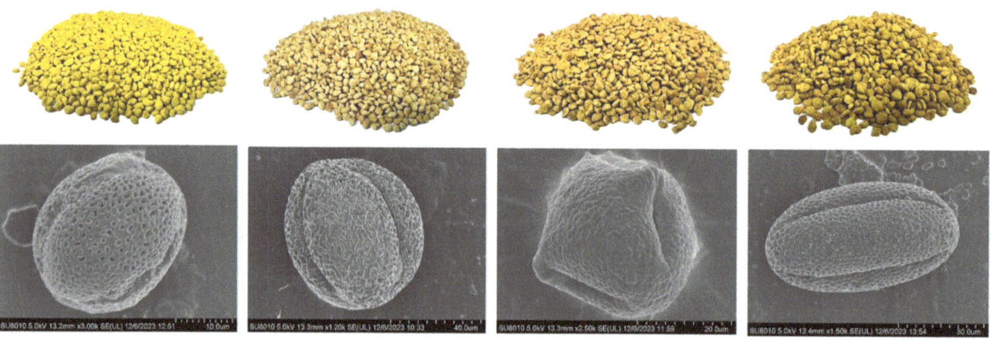

Brassica campestris bee pollen   *Nelumbo nucifera* bee pollen   *Camellia japonica* bee pollen   *Fagopyrum esculentum* bee pollen

Figure 1. Physical and electron microscope images of four varieties of bee pollen.

3.2. E-Nose Analysis of Bee Pollen

PCA operates as an unsupervised method to transform a set of possibly correlated variables into a linearly uncorrelated set of variables via orthogonal transformation. It is a valuable tool for data analysis and feature extraction that can be effectively combined with other pattern recognition algorithms to improve data separability and model performance [17]. PCA is a multidimensional data analysis approach with quantitative variables.

Sample similarity represents small differences, and distance represents a noticeable component difference. PCA is typically employed to reveal the relationships among multiple variables via a few principal components or to extract a few principal components from the original variable while maintaining as much information about the original variable as possible [18]. PC1 is the predominant feature within the multidimensional data matrix, with PC2 following closely behind as the second most significant attribute in the dataset. PLS-DA is a supervised identification method predominantly used to identify the differences between samples of different classes. However, the model cannot identify variables and discard non-informational variables. OPLS-DA is an improved PLS-DA method using distinct projections and orthogonal components to characterize the variation between and within groups. OPLS-DA can eliminate data irrelevant to the category information (orthogonal) by orthogonalization. Additionally, compared to other approaches, it can more easily exclude independent variables unrelated to classification and screen out characteristic variables of samples. OPLS-DA is used to obtain optimal classification and establish discriminant models. OPLS-DA models have been widely utilized in food traceability or screening and the identification of differences in metabolomics [19]. Therefore, we utilized PCA and OPLS-DA to visually illustrate the distinctions in e-nose outcomes across four varieties of bee pollen.

The PCA score plot shown in Figure 2A illustrates the classification of four distinct varieties of bee pollen samples, with PC1 and PC2 accounting for a cumulative contribution rate of 98.331%. This encompasses the valid representation of the sample characteristics. The proximity of samples indicates their similarity, with closer distances reflecting smaller dissimilarities; conversely, greater separations between samples indicate more pronounced differences. The Bc sample is located independently on the left side of the designated area, while the remaining samples are positioned within the right-side region, indicating a significant disparity in overall olfactory characteristics between the Bc sample and its counterparts. There are discernible variations in the olfactory characteristics of the remaining three samples. Furthermore, the OPLS-DA score plot (Figure 2B) indicates distinct separation among the four pollen samples. Notably, the proximity between Nn and Fe in the score plot is comparatively closer than the other pollens, suggesting a lesser disparity between them relative to other pollen varieties. PCA and OPLS-DA revealed pronounced discrimination among the four varieties of bee pollen, indicating substantial variations in odor profiles across the samples.

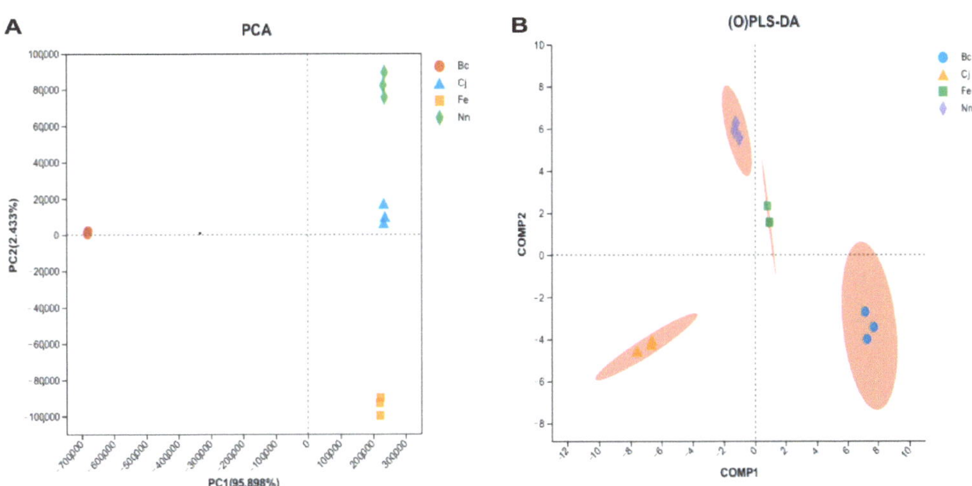

Figure 2. PCA score plot (**A**) and OPLS-DA score plot (**B**) of four varieties of bee pollen according to e-nose analysis.

The gas chromatography plots of distinct bee pollen samples (Figure 3) were produced with Origin version 2022 software. Based on the findings depicted in Figure 3, clear variations are present in the chromatographic data of different varieties of bee pollen. Initially, we utilized PCA and OPLS-DA to rapidly identify components in the samples that exhibit significant variations and contribute significantly to flavor. Subsequently, we employed the Arochembase database to determine volatile odor substances with specific characteristics. The detailed qualitative and quantitative findings are outlined in Table 2. Because the quantitative and qualitative compositions of volatile compounds are primarily associated with floral species and, to a lesser extent, with climatic conditions and geographical locations, each pollen type has a unique volatile compound profile [20]. For instance, the Bc sample contains a diverse range of compounds, including 2-methyl-2-propanol, propan-2-one, 2-propanol, n-butanol, and 2-methylbutanal. Significantly, the predominant presence of n-butanol contributes to the alcoholic, amyl alcohol, and banana-like aroma notes, as well as cheese-like undertones, with fermented and fruity characteristics identified in Bc reaching 740,380, which is approximately 100 times higher than that found in Nn (7689). 2-Methyl-2-propanol characterized the camphor odor of bee pollen from Bc reaching 123,605, which is six times more than that from Nn (19,961).

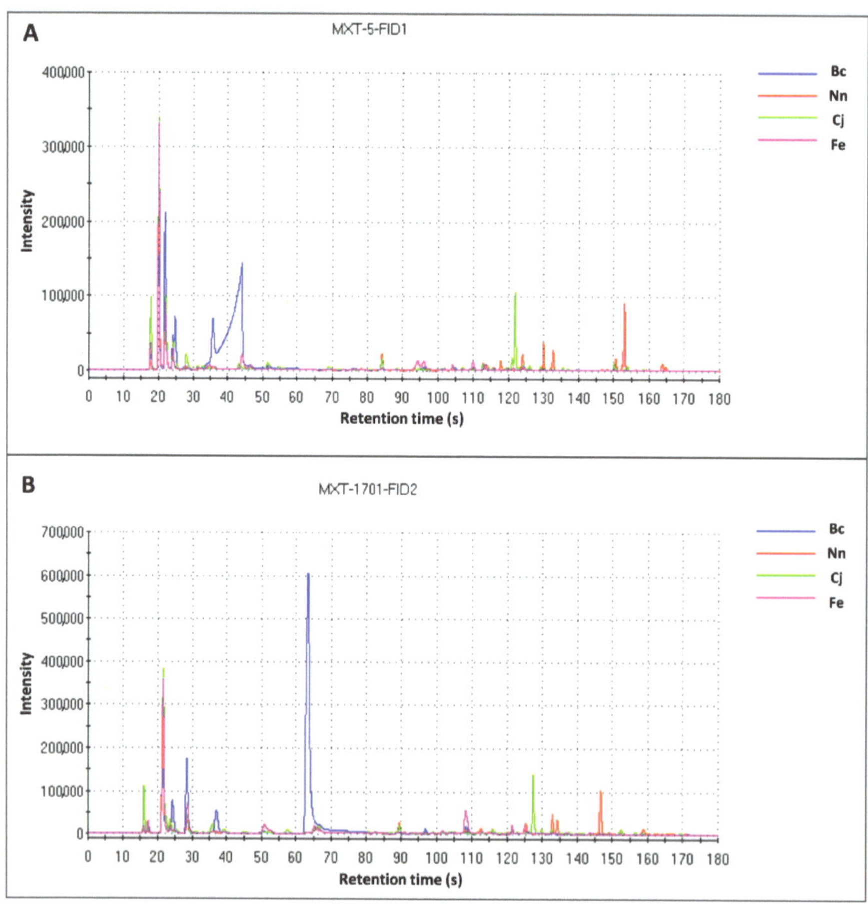

Figure 3. Gas chromatogram of the four varieties of bee pollen collected by MXT-5-FID1 (**A**) and MXT-1701-FID2 (**B**) columns, respectively.

Table 2. Qualitative and relative quantitative results of gas chromatographic data of the four varieties of bee pollen samples.

Compounds	Retention Time -Column 5 (s)	Retention Time -Column 1701 (s)	CAS	Odor Description	Bc	Nn	Cj	Fe
Methyl formate	17.73	17.30	107-31-3	Agreeable; Fruity; Plum	11,469	4321	39,886	15,056
Ethanol	19.90	21.35	64-17-5	Alcoholic; Ethanol; Fragrant; Pleasant; Pungent; Sweet	141,776	165,650	256,029	212,202
2-Methyl-2-propanol	21.83	28.61	75-65-0	Camphor	123,605	19,961	30,873	51,927
Propan-2-one	24.01	23.46	67-64-1	Apple; Characteristic; Fruity; Pear; Solvent; Sweet; Violet	22,363	19,003	20,321	16,956
2-Propanol	24.79	24.32	67-63-0	Acetone; Alcoholic; Ethanol; Floral; Pleasant; Woody	40,371	1391	3655	0
1-Propanol	27.90	35.95	71-23-8	Alcoholic; Ethanol; Fermented; Fruity; Fusel; Plastic; Pungent	3874	2572	16,968	5417
Methyl propanoate	35.82	39.33	554-12-1	Apple; Fresh; Fruity; Rum; Strawberry; Sweet	3861	2370	10,038	3088
(E)-2-Butenal	43.42	57.49	123-73-9	Floral; Plastic; Pungent	1028	1046	8496	264
n-Butanol	44.30	63.51	71-36-3	Alcoholic; Amyl alcohol; Banana; Cheese; Fermented; Fruity; Fusel; Harsh; Medicinal; Oil; Sweet	740,380	7689	4109	26,918
2-Methylbutanal	46.37	51.96	96-17-3	Almond; Apple; Burnt; Cocoa; Coffee; Fermented; Fruity; Iodoform; Malty; Nutty; Sour	6561	850	811	6399
Pent-1-en-3-ol	51.37	66.32	616-25-1	Burnt; Butter; Fruity; Grassy; Horseradish; Meaty; Milky; Pungent; Vegetable	4994	9368	10,060	2565
(E)-2-Pentenal	69.10	80.49	1576-87-0	Apple; Fruity; Green; Oily; Orange; Pungent; Soapy; Strawberry; Tomato	1233	693	4055	2500
Pentanol	72.81	85.78	71-41-0	Alcoholic; Anise; Balsamic; Fruity; Fusel; Oil; Pungent; Sweet; Waxy	227	934	797	258
Hexanal	84.01	89.44	66-25-1	Acorn; Aldehydic; Fatty; Fishy; Fresh; Fruity; Grassy; Herbaceous; Leafy; Sharp; Sweaty; Tallowy; Vinous	2141	13,722	10,314	3073
Ethyl trans-2-butenoate	85.99	95.48	623-70-1	Alliaceous; Chemical; Pungent; Rum; Sweet	1260	602	596	846
Methyl pentanoate	88.45	90.92	624-24-8	Apple; Fruity; Nutty; Pineapple; Sweet	1844	1125	1563	700
E-2-Hexen-1-ol	94.74	102.21	928-95-0	Banana; Butter (cooked); Fresh; Fruity; Leafy; Medicinal; Walnut	2265	792	951	24,072
(Z)-3-Hexen-1-ol	96.11	98.92	928-97-2	Earthy; Floral; Fresh; Fruity; Leafy; Mossy; Oily; Petal	3033	999	1591	21,817
Benzaldehyde	104.01	121.48	100-52-7	Almond; Bitter; Bitter almond; Burnt sugar; Cherry; Fruity; Malty; Oil; Pepper; Sharp; Sweet; Woody	922	999	2759	4799
Sabinene	106.99	105.12	3387-41-5	Citrus; Fresh; Pepper; Pine; Spicy; Sweet; Turpentine; Woody	2592	2066	2318	1863
Amyl propanoate	109.21	112.56	624-54-4	Apricot; Fruity; Pineapple; Sweet	1036	968	989	8780
trans-Hex-2-enyl acetate	110.41	116.65	2497-18-9	Apple; Banana; Fresh; Sweet; Waxy	583	853	1075	0
Octanal	112.63	115.85	124-13-0	Aldehydic; Citrus; Fatty; Floral; Fruity; Lemon; Meat (boiled); Orange; Orange peel; Pungent; Soapy; Stew; Waxy	918	736	412	687
Myrcene	113.77	108.06	123-35-3	Balsamic; Fruity; Geranium; Lemon; Metallic; Plastic; Pleasant; Resinous; Soapy; Spicy; Sweet; Woody	6442	13,005	6597	5918
alpha-Terpinene	115.47	114.86	99-86-5	Citrus; Fruity; Gasoline; Lemon; Medicinal; Woody	2390	3955	2350	1503
Benzeneacetaldehyde	118.39	121.52	122-78-1	Cocoa; Floral; Grassy; Hawthorn; Honey; Hyacinth; Rose; Sweet	976	8042	1403	959

Table 2. *Cont.*

Compounds	Retention Time -Column 5 (s)	Retention Time -Column 1701 (s)	CAS	Odor Description	Bc	Nn	Cj	Fe
(E)-2-Octenal	119.98	122.99	2548-87-0	Burdock; Burnt; Fatty; Fruity; Mushroom; Nutty; Sour; Sweet; Tallowy; Waxy	2179	2319	2561	1538
Linalool	121.78	127.27	78-70-6	Anise; Bergamot; Citrus; Floral; Fragrant; Fresh; Fruity; Lavender; Lemon; Lily; Muscat; Oil; Parsley; Rose; Spicy; Sweet; Terpenic; Woody	1076	2472	47,271	1884
2-Phenylethanol	125.95	129.70	60-12-8	Floral; Flower; Honey; Lilac; Perfumery; Rose; Spicy	3241	14,188	11,858	4346
Camphor	129.99	132.94	76-22-2	Aromatic; Camphor; Fragrant; Leafy	1502	16,620	4886	1520
Methyl salicylate	132.73	134.34	119-36-8	Berry; Minty; Peppermint; Sweet; Winey; Wintergreen	464	12,074	866	586
Decanal	136.85	132.88	112-31-2	Aldehydic; Burnt; Citrus; Fatty; Floral; Herbaceous; Lemon; Orange; Orange peel; Soapy; Stew; Sweet; Tallowy; Waxy	2223	21,070	7094	1239
Ethyl nonanoate	139.29	137.78	123-29-5	Fruity; Rose; Rum; Waxy	1278	1283	1129	796
Dodecanal	146.23	144.73	112-54-9	Aldehydic; Caprylic; Citrus; Fatty; Floral; Herbaceous; Lily; Oily; Soapy; Waxy	783	2162	697	615
beta-Himachalene	150.69	152.62	1461-03-6	-	923	45,690	5719	459
beta-Caryophyllene	153.05	146.61	87-44-5	Fruity; Spicy; Sweet; Terpenic; Woody	254	1329	2735	207
Nonanoic acid hexyl ester	163.75	158.78	6561-39-3	Brandy; Floral; Fruity; Vegetable	274	11,476	789	246
Nonadecane	176.73	170.74	629-92-5	Alkane; Fuel; Fusel	58	391	25	122

Additionally, the Nn sample exhibits significant levels of hexanal, myrcene, alpha-terpinene, benzeneacetaldehyde, 2-phenylethanol, camphor, methyl salicylate, decanal, beta-himachalene, and nonanoic acid hexyl ester. Notably, decanal and beta-himachalene are prominent contributors to the characteristic olfactory profile of Nn, encompassing aldehydic notes alongside smoky nuances and citrusy undertones. Fatty elements alongside floral and herbaceous hints reminiscent of lemon peel and orange zest are also present. The aroma also exhibits soapy aspects with subtle stew-like qualities while maintaining a sweet yet tallowy essence complemented by waxy undertones. In the Cj sample, significant levels of methyl formate, ethanol, 1-propanol, methyl propanoate, (E)-2-butenal, pent-1-en-3-ol, (E)-2-pentenal, linalool, and beta-caryophyllene were found. Among these compounds, ethanol dominates and contributes to the olfactory profile of Cj, characterized by alcoholic notes alongside fragrant and pleasant undertones accompanied by pungency and sweetness. The Fe sample exhibits substantial levels of E-2-hexen-1-ol, (Z)-3-hexen-1-ol, benzaldehyde, and amyl propanoate. Notably, E-2-hexen-1-ol and (Z)-3-hexen-1-ol, with relative contents reaching 24,072 and 21,817, dominate the odor profile of Fe, exceeding those in Nn by 30 and 20 times, respectively, characterized by notes of banana, cooked butter, freshness, fruitiness, leafiness, medicinal qualities, walnut-like nuances, earthy undertones with floral hints, and mossy accents along with an oily petal-like aroma.

Based on research conducted by Bi et al., certain volatile organic compounds, including styrene, limonene, nonanal, and hexanal, are pivotal constituents that contribute to the distinctive aroma of yellow bee pollen [21]. Because of the high protein and lipid levels in bee pollen, exposure to oxygen, heat, or enzymes triggers protein hydrolysis, fat decomposition, and enzyme oxidation. Consequently, these reactions enhance the Maillard reaction and Strecker degradation pathways, causing the formation of distinctive flavor compounds. According to research by Ni, a total of 147 volatile organic compounds were found in the Nn sample, with aldehydes and terpenoids comprising most of these compounds [22]. A total of 42 aldehyde compounds were found, with the highest concentration observed for 2-pental (E) in fresh Nn. Terpene compounds were the predominant volatile constituents in fresh Nn samples. According to research conducted by Cai et al., alcohols make up a significant proportion (69.27%) of the total volatile components found in bee pollen, while aldehydes, ketones, esters, phenolic acids, and sulfides collectively contribute to 9.6% of the overall volatile composition [23]. The concentration of 4.6-dimethyl-dodecane in Bc reached 13.03%. In bee pollen, a total of 40 characteristic aromatic components have been found, including trans-2-nonenic acid, nonanoic acid, 10-undecylenal, beta-cyclocitral, isopentenol, 5-hydroxymethylfural, linalyl acetate, ethyl nonanoate, geranyl propionate, and beta-caryophyllene. These compounds contribute to the development of a luscious creamy flavor, buttery fragrance, floral and fruity aroma, invigorating sensation, and a subtle hint of marine essence. Nakib et al. identified a total of 67 volatile compounds, classified into acids, alcohols, aldehydes, alkanes, aromatic alcohols, benzene derivatives, chromene derivatives, esters, furans, ketones, nitrile, nitrogen compounds, phenols, sulfur compounds, terpenes, and others [24].

3.3. E-Tongue Analysis of Bee Pollen

The response curve of an e-tongue is made up of three distinct stages: the baseline stage, the variation stage, and the stable stage. In our detection, the baseline phase was brief, whereas the change phase duration varied across sensors but typically reached stability within 30 s. The response curves of these seven taste sensors (AHS, PKS, CTS, NMS, CPS ANS, and SCS) can be categorized into three scenarios: (1) a progressive increase in the response signal over time; (2) a relatively stable response signal throughout the observation period; and (3) a gradual decline in the response signal over time. Taking Bc, for example (Figure 4), it notably exhibited a diverse complement of seven distinct tastes, each characterized by varying perceived intensities. Notably, AHS is the taste with the highest intensity, while PKS demonstrates a gradual decline in intensity over time. The results indicate that the e-tongue system has a remarkable capacity to elicit responses to the identified samples.

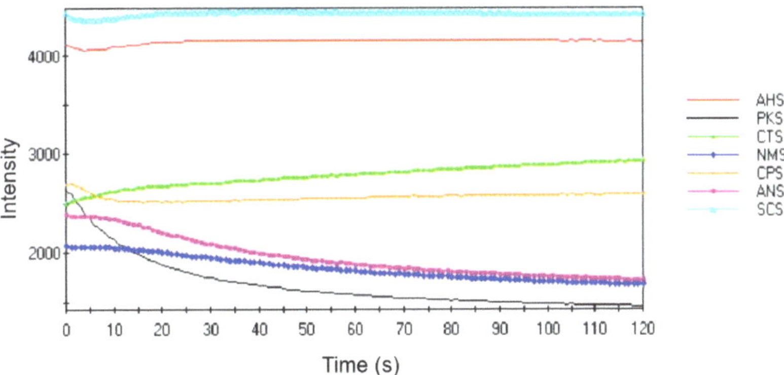

Figure 4. Original signal response graph of the *Brassica campestris* bee pollen (Bc) sample.

Moreover, notable variations in taste exist among different varieties of bee pollen samples. In Figure 5A, it is clear that both Bc and Nn are positioned towards the left side of the confidence interval, while Cj and Fe are situated on the right side. Additionally, there is a discernible disparity in the differentiating impact identified among these four pollen specimens. The cumulative contribution rates of PC1 and PC2 are 99.653%, effectively capturing the true representation of the samples. OPLS-DA analysis also indicated a significant distinction in taste profiles among the four types of bee pollen. The sample tastes exhibit a higher degree of similarity as the distance between them decreases. The higher proximity of Cj and Fe in the scoring chart relative to other pollen types implies a reduced divergence between them compared to other pollen.

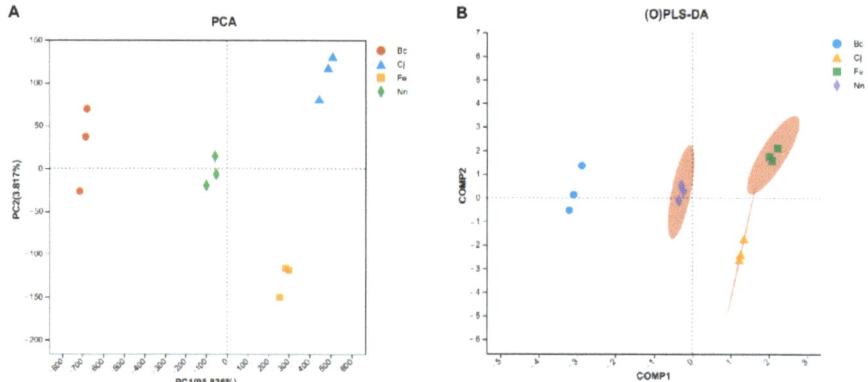

Figure 5. PCA score (**A**) and OPLS-DA score plots (**B**) of four varieties of bee pollen according to e-tongue analysis.

A *t*-test was utilized to characterize the presence of a statistically significant distinction between the two datasets. The fundamental principle involves comparing the mean values of the two datasets, considering the extent of variability and sample size, to ascertain the presence of a statistically significant disparity between them. Based on the *t*-test analysis (Table 3 and Figure 6), notable differences were identified in the taste profiles among the four varieties of bee pollen. The trend of acidity (AHS) and bitterness (SCS) remained consistent and distinct across the four varieties, with the Cj sample exhibiting the most pronounced acidity and bitterness, while the Bc sample demonstrated the lowest expression of both. A noticeable disparity in saltiness (CTS) was observed between Bc and Nn, with the Bc samples exhibiting a comparatively milder saltiness profile, while no statistically significant differentiation was found between Cj and Fe samples. The umami (NMS) response exhibited significant differences between the Fe sample and the remaining three bee pollen varieties, with Fe demonstrating comparatively weaker umami taste. However, no statistically significant difference was observed between the Bc and Cj samples, as well as between the Nn and Cj samples. Notably, a discernible difference was present between the Bc and Nn samples. Regarding sweetness (ANS), significant differences were observed between the Fe samples and the other three varieties of bee pollen, with Fe exhibiting a comparatively lower sweet taste. No statistically significant difference was identified between Bc and Cj, as well as between Bc and Nn. However, there was a notable dissimilarity between Nn and Cj. Moreover, the intensity of AHS and SCS flavors in these four varieties of bee pollen was significantly higher compared to the intensity of other tastes.

Table 3. Significant differences in taste among four varieties of bee pollen, based on e-tongue analysis (*p*-value).

	Bc vs. Nn	Bc vs. Cj	Bc vs. Fe	Nn vs. Cj	Nn vs. Fe	Cj vs. Fe
AHS	0.001 *	0.000	0.001	0.000	0.000	0.000
PKS	0.099	0.021	0.149	0.003	0.012	0.002
CTS	0.020	0.005	0.000	0.000	0.016	0.988
NMS	0.047	0.150	0.008	0.069	0.000	0.003
CPS	0.365	0.216	0.044	0.013	0.003	0.001
ANS	0.142	0.757	0.032	0.023	0.001	0.002
SCS	0.001	0.001	0.001	0.000	0.000	0.000

* *p*-value < 0.05 indicates significant differences between each pair of bee pollen samples.

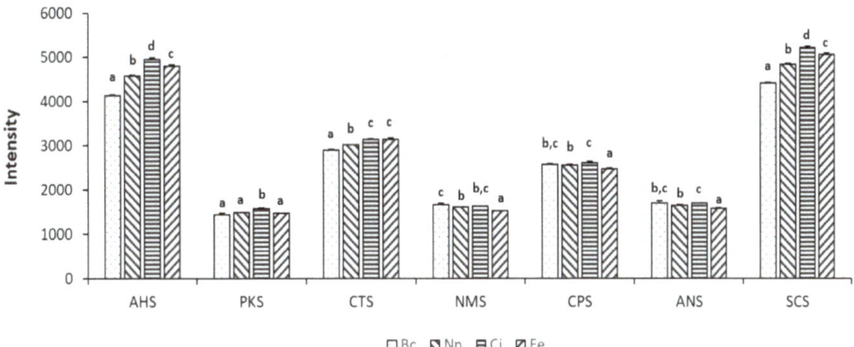

Figure 6. Bar chart illustrating the flavor differences between four varieties of bee pollen analyzed using an e-tongue. Different letter labels indicate significant differences.

The radar chart presented in Figure 7 depicts the taste profiles of four distinct pollen samples, with the values representing the relative intensity of diverse tastes on a scale from 0 to 1. Discernible distinctions exist among the taste profiles of these four varieties of bee pollen. Moreover, this chart can be utilized to evaluate the relative strength of different tastes across various samples. The taste profile of SCS exhibits the highest relative intensity, while PKS has the lowest relative intensity. With respect to the taste intensity of SCS, ANS, CPS, and PKS, the four varieties of bee pollen can be arranged as Bc > Nn > Cj > Fe. The taste intensity of CTS remained consistent across both the Cj and Fe samples, while the Bc samples exhibited a higher intensity relative to the Nn samples. There was no significant variation in taste intensity for PKS and ANS across the four varieties of bee pollen. AHS has a slightly diminished taste profile compared to SCS in the four varieties of bee pollen.

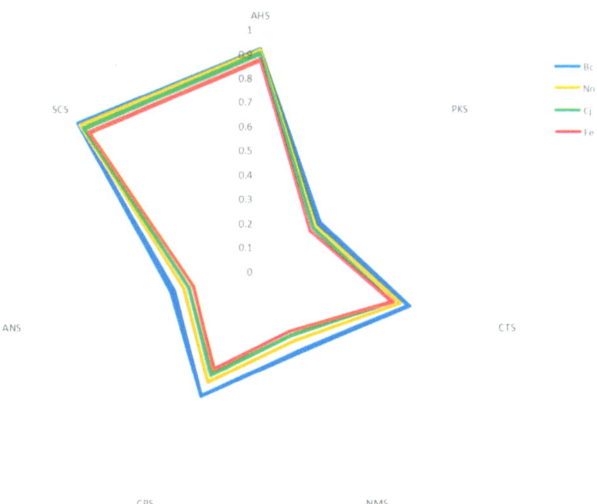

Figure 7. E-tongue radar chart for four varieties of bee pollen.

3.4. Correlation Analysis between E-Nose and E-Tongue Datasets

Correlation analysis was utilized to assess the extent of correlation between two or more variables. As illustrated in Figure 8, each row corresponds to a distinct flavor, while each column represents an individual compound. The red hue signified a positive correlation between flavor and compounds, while a blue shade denoted a negative correlation.

The intensity of color reflects the magnitude of this correlation. Our findings suggest that flavor components exhibited a positive correlation with the majority of esters, aldehydes, and alcohols, while underscoring a negative correlation with hydrocarbons. The flavor of PKS exhibited a positive correlation with linalool levels as well as a negative correlation with n-butanol, 2-methylbutanal, and ethyl trans-2-butenoate concentrations. The flavor of CTS has a positive correlation with benzaldehyde levels. The ANS and NMS levels positively correlate with propan-2-one, 2-propanol, methyl pentanoate and sabinene levels. The flavor of AHS and SCS exhibits a positive correlation with ethanol levels and a negative correlation with octanal concentrations. Furthermore, the correlation analysis findings for flavor components between ANS and NMS and AHS and SCS exhibited a substantial degree of similarity.

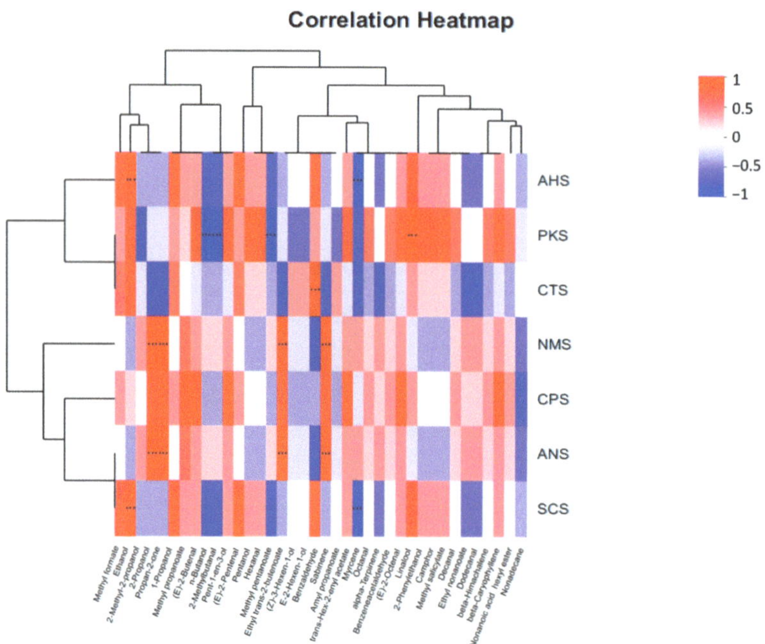

Figure 8. Correlation analysis between e-nose and e-tongue examination of bee pollen. *** *p*-value < 0.001.

4. Conclusions

Flavor chemical analysis employing rapid e-nose and e-tongue techniques unveiled significant variations in both odor and taste among the pollen isolated from four distinct plant species, namely *Brassica campestris* (Bc), *Nelumbo nucifera* (Nn), *Camellia japonica* (Cj), and *Fagopyrum esculentum* (Fe). The analysis performed by the e-nose indicates that Bc, Nn, and Cj contain n-butanol, decanal, and ethanol as their primary odor compounds, respectively. In contrast, Fe predominantly consists of E-2-hexen-1-ol and (Z)-3-hexen-1-ol as its main odorants. Consequently, these substances exhibit distinct characteristics. The e-tongue analysis reveals that bee pollen samples offer a wide range of seven tastes: AHS, PKS, CTS, NMS, CPS, ANS, and SCS. Notably, there is significant variation in taste intensity across various bee pollen samples. By integrating the data acquired from both the e-nose and e-tongue analyses together, the taste components generally exhibited a positive correlation with esters, aldehydes, and alcohols while displaying a negative correlation with hydrocarbons. These findings serve as a theoretical foundation for the comprehensive processing and quality control of bee pollen products.

Author Contributions: Conceptualization, Q.L.; Software, Y.Z.; Validation, E.Z.; Formal analysis, C.L.; Investigation, C.L. and E.Z.; Data curation, Y.Z.; Writing—original draft, C.L.; Writing—review & editing, Q.L.; Supervision, Q.L. and L.W.; Project administration, L.W.; Funding acquisition, L.W. All authors have read and agreed to the published version of the manuscript.

Funding: This work was supported by the Earmarked Fund for CARS (CARS-44), and the Agricultural Science and Technology Innovation Program under Grant (CAAS-ASTIP-2023-IAR).

Institutional Review Board Statement: Not applicable.

Informed Consent Statement: Not applicable.

Data Availability Statement: The original contributions presented in the study are included in the article, further inquiries can be directed to the corresponding authors.

Conflicts of Interest: The authors declare no conflicts of interest.

References

1. Thakur, M.; Nanda, V. Composition and functionality of bee pollen: A review. *Trends Food Sci. Technol.* **2020**, *98*, 82–106. [CrossRef]
2. Giampieri, F.; Quiles, J.L.; Cianciosi, D.; Forbes-Hernández, T.Y.; Orantes-Bermejo, F.J.; Alvarez-Suarez, J.M.; Battino, M. Bee products: An emblematic example of underutilized sources of bioactive compounds. *J. Agric. Food Chem.* **2022**, *70*, 6833–6848. [CrossRef]
3. Wang, S.; Bi, Y.; Zhou, Z.; Peng, W.; Tian, W.; Wang, H.; Fang, X. Effects of pulsed vacuum drying temperature on drying kinetics, physicochemical properties and microstructure of bee pollen. *LWT-Food Sci. Technol.* **2022**, *169*, 113966. [CrossRef]
4. Chen, M.; Yang, X.; Ji, Z.; Zhao, H.; Cheng, N.; Cao, W. Combined treatment of drying, ethanol, and cold plasma for bee pollen: Effects on microbial inactivation and quality attributes. *Food Biosci.* **2024**, *57*, 103542. [CrossRef]
5. Duan, H.; Dong, Z.; Li, H.; Li, W.; Shi, S.; Wang, Q.; Cao, W.; Fang, X.; Fang, A.; Zhai, K. Quality evaluation of bee pollens by chromatographic fingerprint and simultaneous determination of its major bioactive components. *Food Chem. Toxicol.* **2019**, *134*, 110831. [CrossRef]
6. Zhou, W.; Yan, Y.; Mi, J.; Zhang, H.; Lu, L.; Luo, Q.; Li, X.; Zeng, X.; Cao, Y. Simulated digestion and fermentation in vitro by human gut microbiota of polysaccharides from bee collected pollen of Chinese wolfberry. *J. Agric. Food Chem.* **2018**, *66*, 898–907. [CrossRef]
7. Breda, L.S.; de Melo Nascimento, J.E.; Alves, V.; de Toledo, V.D.A.A.; de Lima, V.A.; Felsner, M.L. Green and fast prediction of crude protein contents in bee pollen based on digital images combined with Random Forest algorithm. *Food Res. Int.* **2024**, *179*, 113958. [CrossRef] [PubMed]
8. Lu, L.; Hu, Z.; Hu, X.; Li, D.; Tian, S. Electronic tongue and electronic nose for food quality and safety. *Food Res. Int.* **2022**, *162*, 112214. [CrossRef] [PubMed]
9. Xia, H.; Chen, W.; Hu, D.; Miao, A.; Qiao, X.; Qiu, G.; Liang, J.; Guo, W.; Ma, C. Rapid discrimination of quality grade of black tea based on near-infrared spectroscopy (NIRS), electronic nose (E-nose) and data fusion. *Food Chem.* **2024**, *440*, 138242. [CrossRef]
10. Estivi, L.; Buratti, S.; Fusi, D.; Benedetti, S.; Rodríguez, G.; Brandolini, A.; Hidalgo, A. Alkaloid content and taste profile assessed by electronic tongue of *Lupinus albus* seeds debittered by different methods. *J. Food Compos. Anal.* **2022**, *114*, 104810. [CrossRef]
11. Roy, R.B.; Tudu, B.; Shaw, L.; Jana, A.; Bhattacharyya, N.; Bandyopadhyay, R. Instrumental testing of tea by combining the responses of electronic nose and tongue. *J. Food Eng.* **2012**, *110*, 356–363.
12. Kim, Y.; Lee, U.; Eo, H.J. Effect of NaCl pretreatment on the relationship between the color characteristics and taste of *Cirsium setidens* processed using a micro-oil-sprayed thermal air technique. *Plants* **2023**, *12*, 3193. [CrossRef] [PubMed]
13. Li, Q.; Sun, M.; Wan, Z.; Liang, J.; Betti, M.; Hrynets, Y.; Xue, X.; Wu, L.; Wang, K. Bee pollen extracts modulate serum metabolism in lipopolysaccharide-induced acute lung injury mice with anti-inflammatory effects. *J. Agric. Food Chem.* **2019**, *67*, 7855–7868. [CrossRef]
14. Végh, R.; Csóka, M.; Sörös, C.; Sipos, L. Food safety hazards of bee pollen—A review. *Trends Food Sci. Technol.* **2021**, *114*, 490–509. [CrossRef]
15. Dong, J.; Gao, K.; Wang, K.; Xu, X.; Zhang, H. Cell wall disruption of rape bee pollen treated with combination of protamex hydrolysis and ultrasonication. *Food Res. Int.* **2015**, *75*, 123–130. [CrossRef]
16. Jabeen, S.; Zafar, M.; Ahmad, M.; Ali, M.A.; Elshikh, M.S.; Makhkamov, T.; Mamarakhimov, O.; Yuldashev, A.; Khaydarov, K.; Gafforov, Y.; et al. Micrometer insights into *Nepeta* genus: Pollen micromorphology unveiled. *Micron* **2024**, *177*, 103574. [CrossRef] [PubMed]
17. Zhang, Z.; Zheng, Z.; He, X.; Liu, K.; Debliquy, M.; Zhou, Y.; Zhang, C. Electronic nose based on metal oxide semiconductor sensors for medical diagnosis. *Prog. Nat. Sci. Mater. Int.* **2024**, *in press*.
18. Huang, G.; Liu, T.; Mao, X.; Quan, X.; Sui, S.; Ma, J.; Sun, L.; Li, H.; Shao, Q.; Wang, Y. Insights into the volatile flavor and quality profiles of loquat (*Eriobotrya japonica* Lindl.) during shelf-life via HS-GC-IMS, E-nose, and E-tongue. *Food Chem. X* **2023**, *20*, 100886. [CrossRef]

19. Kang, C.; Zhang, Y.; Zhang, M.; Qi, J.; Zhao, W.; Gu, J.; Guo, W.; Li, Y. Screening of specific quantitative peptides of beef by LC–MS/MS coupled with OPLS-DA. *Food Chem.* **2022**, *387*, 132932. [CrossRef]
20. Filannino, P.; Di Cagno, R.; Gambacorta, G.; Tlais, A.Z.A.; Cantatore, V.; Gobbetti, M. Volatilome and bioaccessible phenolics profiles in lab-scale fermented bee pollen. *Foods* **2021**, *10*, 286. [CrossRef]
21. Bi, Y.; Ni, J.; Xue, X.; Zhou, Z.; Tian, W.; Orsat, V.; Yan, S.; Peng, W.; Fang, X. Effect of different drying methods on the amino acids, α-dicarbonyls and volatile compounds of rape bee pollen. *Food Sci. Hum. Well.* **2024**, *13*, 517–527. [CrossRef]
22. Ni, J.; Bi, Y.; Vidyarthi, S.K.; Xiao, H.; Han, L.; Wang, J.; Fang, X. Non-thermal electrohydrodynamic (EHD) drying improved the volatile organic compounds of lotus bee pollen via HS-GC-IMS and HS-SPME-GC-MS. *LWT-Food Sci. Technol.* **2023**, *176*, 114480. [CrossRef]
23. Cai, Q. *Analysis of Volatile Components and Identification of Main Aroma Components in Four Kinds of Bee Products*; Guangdong Ocean University: Zhanjiang, China, 2023. (In Chinese)
24. Nakib, R.; Rodríguez-Flores, M.S.; Escuredo, O.; Ouelhadj, A.; Coello, M.C.S. Retama sphaerocarpa, Atractylis serratuloides and Eruca sativa honeys from Algeria: Pollen dominance and volatile profiling (HS-SPME/GC–MS). *Microchem. J.* **2022**, *174*, 107088. [CrossRef]

 *foods*

Article

The Theory of Planned Behavior and Antecedents of Attitude toward Bee Propolis Products Using a Structural Equation Model

Kyung-A Sun [1] and Joonho Moon [2],*

1 Department of Tourism Management, Gachon University, Sungnam-si 13120, Republic of Korea; kasun@gachon.ac.kr
2 Department of Tourism Administration, Kangwon National University, Chuncheon 24341, Republic of Korea
* Correspondence: joonhomoon@kangwon.ac.kr

Abstract: This work examines consumers' perceptions of products containing bee propolis using the theory of planned behavior as a theoretical foundation. As antecedents of attitude, this work employs price fairness, healthiness, eco-friendliness, and ease of use. A survey was issued to participants who had experience using bee propolis products and who were recruited using the Clickworker platform service. In total, 305 valid observations were collected for analysis. This study used a maximum likelihood-based structural equation model to test the research hypotheses and find that price fairness, healthiness, eco-friendliness, and ease of use positively affected attitude. Moreover, the intention to use is positively impacted by attitude, subjective norms, and behavioral control. This research contributes to the literature by demonstrating the explanatory power of the theory of planned behavior with respect to bee propolis products.

Keywords: bee propolis products; attitude; theory of planned behavior; price fairness; healthiness; eco-friendliness; ease of use

Citation: Sun, K.-A.; Moon, J. The Theory of Planned Behavior and Antecedents of Attitude toward Bee Propolis Products Using a Structural Equation Model. *Foods* **2024**, *13*, 3002. https://doi.org/10.3390/foods13183002

Academic Editor: Jean-Xavier Guinard

Received: 20 August 2024
Revised: 11 September 2024
Accepted: 21 September 2024
Published: 22 September 2024

1. Introduction

Bee propolis products (BPPs) in the United States had a market size of approximately 636 million US dollars in 2021, and these products are projected to achieve a market growth rate of 2.8% between 2022 and 2028 [1]. Bee propolis is commonly used for therapeutic purposes [2] and may promote health via its antioxidant effects and ability to minimize the risk of cardiovascular disease [3,4]. Grand View Research [1] also stated that BPP is attractive to consumers because consumers value their health more, and BPP can meet such needs because it has preventive and remedy functions. Thus, bee propolis has the potential to become a particularly lucrative commodity. Under this condition, it is essential to understand consumer characteristics because such information could become the starting point for the market growth of BPPs. However, consumer behavior concerning this product remains largely unexplored.

This research employs the theory of planned behavior, which has been adopted in various domains to investigate consumer behavior [5–8], as the theoretical foundation. The main attributes of this work include attitude, subjective norm, and behavioral control in the theory of planned behavior, and the dependent variable is the intention to use, in line with the previous works [9,10]. Although numerous studies demonstrated the explanatory power of the theory of planned behavior in the consumer behavior domain [11–13], consumers' behavior in the case of BPPs has been sparsely explored using the theory of planned behavior as the theoretical foundation. Considering such a research gap, this research tests the explanatory power of the theory of planned behavior in the case of BPPs.

Attitude is the long-term appraisal of goods and services [5,8,14]. The merits of BPPs are likely to include their affordability [15], indicating that price is likely to become the strength of BPPs. Scholars also argued that healthy properties are the motivation for BPP consumption because it improves an individual's immune system [16,17]. Moreover, prior works noted that eco-friendliness is an appealing point of BPPs because consumers are skeptical of chemical-based medicine [18,19]. Last, researchers claimed that ease of use is an important aspect of consumer choice because complexity causes negative consequences in the area of consumer behavior [20,21]. All things considered, this research employs four attributes to account for the attitude of consumers in the area of BPPs.

In summary, the first objective of this work is to test the explanatory power of the theory of planned behavior in the area of BPPs. The second objective of this research is to investigate the antecedents of attitude, which is a main variable in the theory of planned behavior. As antecedents, this research proposes four attributes: price fairness, healthiness, eco-friendliness, and ease of use. Even though numerous studies have explored BPP characteristics [16,22,23], researchers have insufficiently examined the consumer perception of BPPs. Such a research gap leads this study to inspect consumer perception of BPPs using the theory of planned behavior as a theoretical foundation. Also, this research sheds light on the literature by scrutinizing the determinants of attitude in BPPs. Additionally, this research discusses managerial implications.

2. Review of Literature and Hypothesis Development

2.1. Theory of Planned Behavior

The theory of planned behavior discusses the determinants of intention, including attitude, subjective norms, and behavioral control [7,14]. Attitude is the individual appraisal of a subject based on a long time [6,24]; subjective norms constitute others' expectations of and perspectives on certain behaviors that create social pressure [5,8]. Behavioral control relates to constraints regarding available resources [12,13]. Many studies have used the theory of planned behavior in diverse domains. For instance, Tama et al. [11] explored farmers' intentions regarding the conservation of agriculture using this theory as a theoretical background; Fan et al. [6] used it to examine student behavior concerning vaccinations in China; Ateş [5] investigated the antecedents of pro-environmental behavioral intentions using the theory as the framework; Conner et al. [25] applied it to investigate healthy eating habits; and Chen [26] and Lim and An [27] adopted it as a theoretical foundation to account for consumers' healthier food choices.

2.2. Hypothesis Development and Consumer Research on BPPs

The price of BPPs may support a positive consumer perception of them on the market, and consumers are likely to perceive the price reasonably [15]. Still, their health-promoting properties [16,17], eco-friendliness (they are largely free from chemical contaminants) [18,19], and ease of use (being available in pill, spray, and gel forms) [20,21] are also key strengths. This research thus examines the determinants of attitude using four attributes: price fairness, healthiness, eco-friendliness, and ease of use.

The first area is price fairness. Price fairness refers to how rationally consumers assess the listed price of the seller [28,29]. Price rationality plays a significant role in persuading consumers to purchase a product [28,30]. Rai [29] demonstrated a positive effect of price fairness on attitude. Sohaib et al. [31] examined well-being products that consumers favor and detected a positive association between price fairness and attitude. Syah et al. [15] explored consumers in Brazil, and the findings implied that price significantly affected consumer perception. Such findings are likely to be applied to the case of BPPs because BPPs aim to promote the well-being of consumers.

Perceived healthiness plays a significant role in building positive attitudes and promoting purchase intention [32,33]. Jang et al. [34] found that healthiness positively impacted consumer attitudes toward in-flight meals, while Janssen and Bogaert [35] demonstrated its positive effects based on food packaging information. El-Sakhawy et al. [23] additionally alleged that consumers value BPPs more because they promote individuals' health condition by strengthening the immune system. Based on the review of the literature, the following research hypotheses are proposed:

Hypothesis 1: *Price fairness positively affects attitude.*

Hypothesis 2: *Healthiness positively affects attitude.*

Eco-friendliness refers to a product's environmental impact in terms of its ingredients and the minimization of harmful effects on the environment [36,37]. Given that consumers generally perceive eco-friendly products as safer to eat [37,38], they are more likely to develop positive attitudes toward them. A meta-analysis found that eco-friendliness helps determine attitude [39]. Other studies have reported a positive influence of eco-friendliness on consumer attitude [36,38] because eco-friendly products can make consumers perceive food products more safely. Additionally, Mountford-McAuley et al. [40] suggested that the environmental aspect of BPPs offers a key marketing opportunity, as consumers place greater importance on environmental attributes, particularly in the context of medical products.

Regarding ease of use, which refers to the simplicity with which goods may be used by consumers [41,42], consumers tend to avoid products that they perceive as complex and confusing or that require time and effort to use [43,44]. They may perceive such goods as requiring an extra investment to learn how to use the products, which functions as a deterrent. Previous research has demonstrated a positive association between ease of use and attitude in various areas: e-commerce [45], smartphone chatbots [41], augmented reality [44], and online learning [43]. In the case of functional food such as BPPs, complexity causes negative perception because misuse is likely to be harmful to health conditions. Additionally, Grand View Research [1] also noted that BPPs are likely to be used in various types: ointment, spray, and pill for the convenience of consumers. Hence, we propose the following:

Hypothesis 3: *Eco-friendliness positively affects attitude.*

Hypothesis 4: *Ease of use positively affects attitude.*

Several previous studies have considered the intention to use the consumer behavior domain in the framework of the theory of planned behavior [46,47]. For example, a meta-analysis revealed that attitude, subjective norms, and behavioral control positively affect individual intention [7], with studies showing the positive influences of these on farmers' intentions regarding conservation agriculture [11], intention to use the Alipay e-wallet system [10], and intention to use electronic vehicles [9]. It can be inferred that the attributes in the theory of planned behavior have played a significant role in explaining consumer intention to use. Thus, we propose the following to ensure its accountability in the domain of BPPs:

Hypothesis 5: *Attitude positively affects intention to use.*

Hypothesis 6: *Subjective norms positively affect intention to use.*

Hypothesis 7: *Behavioral control positively affects intention to use.*

3. Method
3.1. Research Model

Figure 1 illustrates the research model, which includes eight attributes: price fairness, healthiness, eco-friendliness, ease of use, attitude, subjective norms, behavioral control, and intention to use. Price fairness, healthiness, eco-friendliness, and ease of use are the determinants of attitude, all of which have a positive impact on attitude. Moreover, intention to use is positively influenced by attitude, subjective norms, and behavioral control.

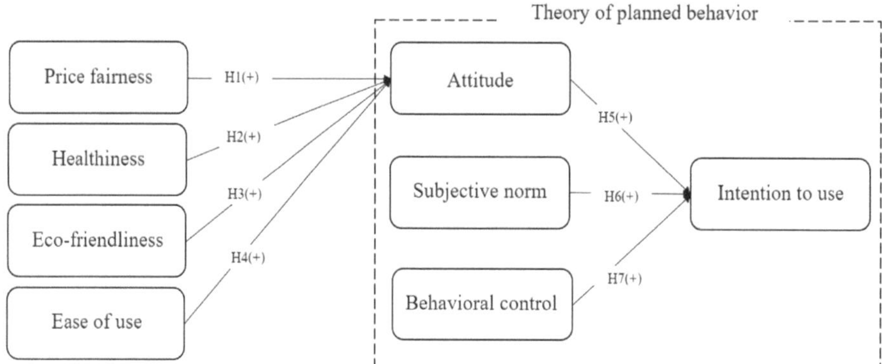

Figure 1. Research model.

3.2. Description of Measurement Items

Table 1 presents the measurement items. A Likert five-point scale (1 = strongly disagree, 5 = strongly agree) was used to measure most attributes, while attitude was measured using a semantic differential scale (e.g., 1 = bad, 5 = good). Price fairness is defined as consumers' assessment of the price of BPPs as being rational. We pull these items from prior studies but modify them for greater applicability to this present study. The main attributes are price fairness [28,31], healthiness [32,34], eco-friendliness [34,36], ease of use [41,44], attitude [48,49], subjective norm [12,13], behavioral control [5,6], and intention to use [9,10,50]. This research defined price fairness as how consumers perceived the price of BPPs. The definitions are given above. The operational definition of healthiness is how consumers perceive the BPPs to promote health conditions. Eco-friendliness is defined as how the BPPs are related to the environmental aspects. Ease of use is defined as how BPP is simple to use. Attitude is defined as a long-term evaluation of BPPs. The definition of the subjective norm is a sort of popularity of BPPs from people. Behavioral control is defined as whether an individual is disturbed by the consumption of BPPs. Last, the operational definition of intention to use is how willing individuals are to purchase BPPs.

Table 1. Description of measurements.

Construct	Code	Item
Price fairness	PF1	The price of BPPs is fair.
	PF2	The price of BPPs is reasonable.
	PF3	The price of BPPs is rational.
	PF4	The price of BPPs is acceptable.
Healthiness	HE1	BPPs support my good health.
	HE2	BPPs are useful for better health.
	HE3	BPPs improve my health condition.
	HE4	BPPs are effective for enhancing my health.
Eco-friendliness	EF1	BPPs are eco-friendly.
	EF2	BPPs are environmentally friendly.
	EF3	BPPs do not cause ecological harm.
	EF4	BPPs are environmentally safe.
Ease of use	EU1	BPPs are easy to use.
	EU2	BPPs are simple to use.
	EU3	BPPs are straightforward to use.
	EU4	It is not complex to use BPPs.

Table 1. *Cont.*

Construct	Code	Item
Attitude	AT1	For me, BPPs are (bad/good).
	AT2	For me, BPPs are (negative/positive).
	AT3	For me, BPPs are (unfavorable/favorable).
	AT4	For me, BPPs are (foolish/wise).
Subjective norm	SN1	People around me seem to consider the use of BPPs naturally.
	SN2	People around me believe that BPPs represent ethical consumption.
	SN3	People close to me believe that the consumption of BPPs is easy.
	SN4	People who are important to me consider it possible to use BPPs.
Behavioral control	BC1	I have enough resources to buy BPPs.
	BC2	I have enough money to buy BPPs.
	BC3	There are no obstacles to my use of BPPs.
	BC4	I have sufficient resources to purchase BPPs.
Intention to use	IU1	I intend to use BPPs.
	IU2	I will purchase BPPs.
	IU3	I am willing to buy BPPs.
	IU4	I have an intention to use BPPs.

3.3. Data Collection

The Clickworker (https://www.clickworker.com/, accessed on 11 June 2024) platform was used to recruit survey participants. Native English speakers were targeted and the data were collected between 23 June and 2 August 2024. Participants were asked whether they had experience using BPPs, and 650 responses were initially collected. Those with no experience using BPPs (345 respondents) were eliminated from the dataset because this research targeted the responses from the vivid BPP experience. Consequently, 305 observations remained for analysis. Table 2 details the survey participants' profiles.

Table 2. Profiles of survey participants (N = 305).

Item	Frequency	Percentage
Male	126	41.3
Female	179	58.7
20–29 years	47	24.3
30–39 years	101	33.1
40–49 years	92	30.2
50–59 years	33	10.8
Older than 60 years	5	1.6
Monthly household income		
Less than USD 2500	86	28.2
Between USD 2500 and USD 4999	111	36.4
Between USD 5000 and USD 7499	48	15.7
Between USD 7500 and USD 9999	13	4.3
More than USD 10,000	47	15.4
Weekly use frequency		
None	118	38.7
One to two times	121	39.7
Three to six times	44	14.4
More than seven times	22	7.2

3.4. Data Analysis

Frequency analysis was used to derive the survey participants' demographic information. Confirmatory factor analysis was performed to examine convergent validity. The convergent validity of measurements was ensured by multiple indices: loading > 0.5, average value extracted (AVE) > 0.5, and construct reliability (CR) > 0.7 [51,52]. The goodness-of-fit was confirmed using multiple indices: Q (CMIN/degrees of freedom) < 3, the goodness-of-fit index (GFI), the normed fit index (NFI), the relative fit index (RFI), the comparative fit index (CFI) > 0.8, and the root mean square error of approximation (RMSEA) < 0.1 [52,53]. Then, the mean values and standard deviations (SDs) were computed for the variables. The correlation matrix was adopted not only to explore the relationships between attributes but also to ensure discriminant validity. The square root of AVE should be greater than the correlation coefficient for an acceptable discriminant validity level [51–53]. A maximum likelihood-based structural equation model was further conducted to test hypotheses using a confidence interval of 90 percent as the threshold.

4. Results

4.1. Convergent Validity and Discriminant Validity

Table 3 presents the results of the confirmatory factor analysis. The goodness-of-fit indices indicate that the results are statistically acceptable (χ^2 = 912.863, df = 442, χ^2/df = 2.065, GFI = 0.837, NFI = 0.899, RFI = 0.886, CFI = 0.945, and RMSEA = 0.059). All factor loadings, AVEs, and CRs were greater than the cut-off value, suggesting that the convergent validity of the measurement items was adequate.

Table 3. Confirmatory factor analysis results.

Construct	Code	Loading	Mean (SD)	AVE	CR
Price fairness	PF1	0.860	3.57 (0.84)	0.724	0.919
	PF2	0.890			
	PF3	0.845			
	PF4	0.807			
Healthiness	HE1	0.885	3.84 (0.86)	0.743	0.920
	HE2	0.868			
	HE3	0.863			
	HE4	0.832			
Eco-friendliness	EF1	0.882	3.92 (0.84)	0.748	0.922
	EF2	0.817			
	EF3	0.895			
	EF4	0.864			
Ease of use	EU1	0.773	4.19 (0.76)	0.734	0.916
	EU2	0.871			
	EU3	0.893			
	EU4	0.885			
Attitude	AT1	0.841	4.00 (0.74)	0.664	0.887
	AT2	0.831			
	AT3	0.881			
	AT4	0.695			
Subjective norm	SN1	0.830	3.55 (0.90)	0.687	0.898
	SN2	0.845			
	SN3	0.843			
	SN4	0.799			
Behavioral control	BC1	0.798	3.66 (0.94)	0.693	0.900
	BC2	0.827			
	BC3	0.823			
	BC4	0.881			
Intention to use	IU1	0.927	3.84 (1.01)	0.833	0.953
	IU2	0.936			
	IU3	0.896			
	IU4	0.899			

Note: SD, standard deviation, goodness-of-fit indices: χ^2 = 912.863, df = 442, χ^2/df = 2.065 GFI = 0.837; NFI = 0.899; RFI = 0.886; CFI = 0.945; RMSEA = 0.059; CR, construct reliability; AVE, average variance extracted.

Table 4 presents the correlation matrix. The diagonal values are greater than the off-diagonal values, indicating that the discriminant validity of the measurement items was appropriate. Intention to use is positively correlated with attitude, subjective norm, behavioral control, price fairness, healthiness, ease of use, and eco-friendliness. Attitude also positively correlates with subjective norms, behavioral control, price fairness, healthiness, ease of use, and eco-friendliness.

Table 4. Correlation matrix.

	1	2	3	4	5	6	7	8
1. Intention to use	0.914							
2. Attitude	0.656 *	0.814						
3. Subjective norm	0.574 *	0.490 *	0.828					
4. Behavioral control	0.432 *	0.365 *	0.499 *	0.832				
5. Price fairness	0.468 *	0.478 *	0.560 *	0.461 *	0.850			
6. Healthiness	0.684 *	0.631 *	0.562 *	0.354 *	0.435 *	0.861		
7. Ease of use	0.533 *	0.505 *	0.416 *	0.467 *	0.350 *	0.493 *	0.856	
8. Eco-friendliness	0.655 *	0.579 *	0.570 *	0.386 *	0.505 *	0.584 *	0.493 *	0.864

Note: * $p < 0.05$; diagonal is the square root of average variance extracted; SD, standard deviation.

4.2. Hypotheses Testing Using Structural Equation Model

Table 5 details the results of the hypothesis testing. Attitude is positively affected by price fairness, healthiness, eco-friendliness, and ease of use. Intention to use is also positively influenced by attitude, subjective norm, and behavioral control. Thus, all hypotheses are supported.

Table 5. Results of hypotheses testing.

Path	β	Critical Ratio	*p* Value	Results
Price fairness → Attitude	0.132 **	2.72	0.006	H1 supported
Healthiness → Attitude	0.345 **	6.37	0.000	H2 supported
Eco-friendliness → Attitude	0.170 **	3.01	0.003	H3 supported
Ease of use → Attitude	0.168 **	2.97	0.003	H4 supported
Attitude → Intention to use	0.787 **	9.58	0.000	H5 supported
Subjective norm → Intention to use	0.325 **	4.87	0.000	H6 supported
Behavioral control → Intention to use	0.106 *	1.75	0.079	H7 marginally supported

Note: * $p < 0.1$, ** $p < 0.05$.

5. Discussion

The first objective of this study was to evaluate the explanatory power of the theory of planned behavior in the context of BPPs. To achieve this, the research focused on four key attributes: attitude, subjective norms, perceived behavioral control, and intention to use. Specifically, the theory of planned behavior was used to assess consumer behavior related to BPPs. Our findings revealed that attitude, subjective norms, and perceived behavioral control had positive effects on the intention to use BPPs. In addition, the results indicated that subjective norm and behavioral control positively affected the intention to use BPPs. Furthermore, attitude showed the strongest impact on the intention to use BPPs compared to subjective norms and behavioral control. The findings of this work are aligned with Conner et al. [25]'s argument that the theory of planned behavior could become a key framework for health-related research areas.

The second objective was to explore the determinants of consumer attitudes toward BPPs. The results are significant in identifying strong correlations between attitude and four key factors: price fairness, healthiness, eco-friendliness, and ease of use. Notably, price fairness positively influenced consumer attitudes, suggesting that BPPs' pricing enhances their market appeal. Namely, the price is an imperative attribute for consumer appraisal, as scholars claimed [28,29]. Additionally, healthiness, eco-friendliness, and ease of use played crucial roles in shaping positive attitudes toward BPPs, indicating that these attributes should be emphasized in marketing strategies. In other words, consumers developed favorable attitudes toward BPPs because they perceived them as promoting personal health, minimizing environmental harm, and being easy to use. Among these factors, healthiness had the strongest impact. Because BPPs aim to promote health by strengthening the immune system, these findings may align with the main motivators underpinning BPP consumption from the extant literature [16,22,23]. Moreover, it can be inferred that the environmental aspect has become more important for appraisals of consumers in the food market, as prior works stated [40]. Furthermore, this research demonstrated the importance of ease of use in the case of functional food by following the reasons that complexity causes cost from the perspective of consumers [43,44].

6. Conclusions

The key theoretical contributions are as follows. First, this research demonstrated the applicability of the theory of planned behavior in the area of BPPs. Although numerous studies have addressed the characteristics of BPPs [16,22,23], few have explored consumer perceptions of BPPs using the theory of planned behavior. Conner et al. [25] highlighted that the theory of planned behavior primarily focuses on healthy eating, suggesting that it is a suitable theoretical framework for explaining consumer behavior related to BPPs. This present study addressed this gap by validating the theory's explanatory power and identifying the key determinants of consumer attitudes toward BPPs. Our results demonstrated the significant effects of price fairness, healthiness, eco-friendliness, and ease of use in consumer research. These findings contribute to the literature by providing a deeper understanding of consumer characteristics in the context of BPPs. Specifically, the results align with existing studies by highlighting the significant influence of price [15] and healthiness [34] in the food product sector. Additionally, they support the findings of previous research, suggesting that eco-friendliness is a key factor in consumer marketing [39]. It can be inferred that eco-friendliness has become increasingly important in consumer behavior due to a higher interest in environmental issues such as global warming. Moreover, the findings reinforce the importance of ease of use, particularly in the functional food domain [44,45].

Several managerial implications emerged. Above all, managers may need to adopt a more conservative approach to price changes to avoid undermining consumers' perceptions of price fairness because varied prices are likely to undermine price fairness perceptions. Additionally, managers may need to implement a price comparison system to provide consumers with greater transparency. Investing in healthier ingredients is also crucial, as healthiness has emerged as a key motivator. Managers could further appeal to consumers by making nutrition labels more visible. Emphasizing both the environmental benefits and ease of use of BPPs is important for fostering positive consumer attitudes. Specifically, marketing messages should highlight information on price, health-related ingredients, eco-friendly features, and product simplicity, as these factors are strongly associated with positive attitudes. Managers may also consider incorporating visuals or messages related to healthiness and eco-friendliness, which are likely to enhance consumer attitudes. Moreover, allocating resources toward convenient product designs, such as ointments, pills, and sprays, could improve the sales of BPPs, given that ease of use is a strong driver of purchase decisions. In addition, resources should be invested in fostering more positive attitudes, given that attitude functions as the antecedent of intention to use. It may also be worthwhile to focus more on peer group marketing, identifying a target market of consumers who possess sufficient resources to consume BPPs. This could ultimately increase the sales of BPPs. Managers also might need to contemplate that focusing on the healthiness-related area for marketing could become the most efficient resource allocation because it shows the strongest effect as compared to the other three attributes.

This research had several limitations that should be mentioned. First, we focused exclusively on the linear effect between attributes, and future research should consider more diverse relationships, such as moderating effects and curvilinear impacts, to further elucidate consumer behavior. Moreover, a survey was used as the sole data-collection instrument; future studies might consider a broader range of methods such as an experimental design. In addition, the study participants were exclusively native English speakers; future works may consider more diverse geographic cases, given that food and nutrients may be perceived differently in various cultural contexts and under the influence of market trends. This research has another limitation because the survey participants were experienced with the BPPs. Future research might be able to consider respondents without experience with BPPs because the outcome might yield further information for the potential market.

Author Contributions: K.-A.S.: formal analysis, writing—original draft; J.M.: writing—review and editing. All authors have read and agreed to the published version of the manuscript.

Funding: This research received no external funding.

Institutional Review Board Statement: According to the exemption standard of Kangwon National University, ethical review and approval were waived for this study due to this research not collecting any personal information (https://irb.kangwon.ac.kr:461/02_board/board03.htm?Item=board3&mode=view&No=103, accessed on 25 July 2024).

Informed Consent Statement: Informed consent was obtained from all subjects involved in this study.

Data Availability Statement: The data presented in this study are available upon request from the corresponding author. The data are not publicly available due to privacy.

Conflicts of Interest: The authors declare no conflicts of interest.

References

1. Grand View Research. Propolis Market Size, Share & Trends Analysis Report by Product Type (Capsules and Tablets, Liquids), by Distribution Channel (Retail Store, Online), by Region (North America, Europe, APAC, CSA, MEA), and Segment Forecasts, 2022–2028. 2024. Available online: https://www.grandviewresearch.com/industry-analysis/propolis-market-report (accessed on 15 June 2024).
2. Kowalczuk, I.; Gębski, J.; Stangierska, D.; Szymańska, A. Determinants of Honey and other bee products use for culinary, cosmetic, and medical purposes. *Nutrients* **2023**, *15*, 737. [CrossRef] [PubMed]
3. Hadi, A.; Rafie, N.; Arab, A. Bee products consumption and cardiovascular diseases risk factors: A systematic review of interventional studies. *Int. J. Food Prop.* **2021**, *24*, 115–128. [CrossRef]
4. Rozman, A.S.; Hashim, N.; Maringgal, B.; Abdan, K. A comprehensive review of stingless bee products: Phytochemical composition and beneficial properties of honey, propolis, and pollen. *Appl. Sci.* **2022**, *12*, 6370. [CrossRef]
5. Ateş, H. Merging theory of planned behavior and value identity personal norm model to explain pro-environmental behaviors. *Sust. Prod. Cons.* **2020**, *24*, 169–180. [CrossRef]
6. Fan, C.; Chen, I.; Ko, N.; Yen, C.; Lin, C.; Griffiths, M.; Pakpour, A. Extended theory of planned behavior in explaining the intention to COVID-19 vaccination uptake among mainland Chinese university students: An online survey study. *Hum. Vaccines Immunother.* **2021**, *17*, 3413–3420. [CrossRef] [PubMed]
7. Hagger, M.S.; Cheung, M.; Ajzen, I.; Hamilton, K. Perceived behavioral controlmoderating effects in the theory of planned behavior: A meta-analysis. *Health Psychol.* **2022**, *41*, 155. [CrossRef] [PubMed]
8. Mohr, S.; Kühl, R. Acceptance of artificial intelligence in German agriculture: An application of the technology acceptance model and the theory of planned behavior. *Precis. Agric.* **2021**, *22*, 1816–1844. [CrossRef]
9. Shalender, K.; Sharma, N. Using extended theory of planned behaviour (TPB) to predict adoption intention of electric vehicles in India. *Environ. Dev. Sust.* **2021**, *23*, 665–681. [CrossRef]
10. Tian, Y.; Chan, T.; Suki, N.; Kasim, M. Moderating Role of Perceived Trust and Perceived Service Quality on Consumers' Use Behavior of Alipay e-wallet System: The Perspectives of Technology Acceptance Model and Theory of Planned Behavior. *Hum. Beh Emer. Tech.* **2023**, *2023*, 5276406. [CrossRef]
11. Tama, R.; Ying, L.; Yu, M.; Hoque, M.; Adnan, K.; Sarker, S. Assessing farmers' intention towards conservation agriculture by using the Extended Theory of Planned Behavior. *J. Environ. Manag.* **2021**, *280*, 111654. [CrossRef]
12. Soliman, M. Extending the theory of planned behavior to predict tourism destination revisit intention. *Int. J. Hosp. Tour. Adm.* **2021**, *22*, 524–549. [CrossRef]

13. Su, Y.; Zhu, Z.; Chen, J.; Jin, Y.; Wang, T.; Lin, C.; Xu, D. Factors influencing entrepreneurial intention of university students in China: Integrating the perceived university support and theory of planned behavior. *Sustainability* **2021**, *13*, 4519. [CrossRef]
14. Ajzen, I. 2020. The theory of planned behavior: Frequently asked questions. *Hum Beh Emer Tech.* **2020**, *2*, 314–324. [CrossRef]
15. Syah, A.; Kumara, D.; Nizar, N.; Hasanah, U. The Effect of Propolis Product Price and Promotion on Consumer Purchase Interest at PT. Nano Herbaltama International in Setu District, South Tangerang City. *Kontigensi J. Ilm. Manaj.* **2022**, *10*, 403–407. [CrossRef]
16. Hwu, Y.; Lin, F.Y. Effectiveness of propolis on oral health: A meta-analysis. *J. Nurs. Res.* **2014**, *22*, 221–230. [CrossRef]
17. Irigoiti, Y.; Navarro, A.; Yamul, D.; Libonatti, C.; Tabera, A.; Basualdo, M. The use of propolis as a functional food ingredient: A review. *Tren. Food Sci. Tech.* **2021**, *115*, 297–306. [CrossRef]
18. Yaman Turan, N.; Turker, E.; Insaatci, Ö. Microparticles loaded with propolis to make antibacterial cotton. *Cell* **2021**, *28*, 4469–4483. [CrossRef]
19. Singh, L.B. Cosmetic and Medicinal Use of Bee's Honey: A Review. *Mon. Peer Rev. Mag. Agri Alli Sci.* **2022**, *1*, 79–84.
20. Kalia, A.; Morya, S.; Neumann, A. Health from the Hive: Therapeutic Potential of Propolis—A Review. *J. Food Bioa.* **2022**, *18*, 77–84. [CrossRef]
21. Maroof, K.; Lee, R.; Siow, L.; Goh, B.; Chen, K.; Gan, S. A new stable and bioactive formulation of *Geniotrigona thoracia* propolis microemulsion for oral delivery. *Food Chem. Adv.* **2023**, *3*, 100514. [CrossRef]
22. Baysan, U.; Elmas, F.; Koç, M. The effect of spray drying conditions on physicochemical properties of encapsulated propolis powder. *J. Food Proc. Engi.* **2019**, *42*, e13024. [CrossRef]
23. El-Sakhawy, M.; Salama, A.; Mohamed, S. Propolis applications in food industries and packaging. *Biomass Convers. Biorefinery* **2023**, *14*, 13731–13746. [CrossRef]
24. Amoako, G.K.; Dzogbenuku, R.; Abubakari, A. Do green knowledge and attitude influence the youth's green purchasing? Theory of planned behavior. *Int. J. Prod. Perf. Mgt.* **2020**, *69*, 1609–1626. [CrossRef]
25. Conner, M.; Norman, P.; Bell, R. The theory of planned behavior and healthy eating. *Health Psychol.* **2002**, *21*, 194. [CrossRef]
26. Chen, M.F. Modeling an extended theory of planned behavior model to predict intention to take precautions to avoid consuming food with additives. *Food Qual. Pref.* **2017**, *58*, 24–33. [CrossRef]
27. Lim, H.R.; An, S. Intention to purchase wellbeing food among Korean consumers: An application of the Theory of Planned Behavior. *Food Qual. Pref.* **2021**, *88*, 104101. [CrossRef] [PubMed]
28. Park, J.; Park, S.; Kim, G. The effect of service price framing in an augmented product on price fairness, service and product attitude. *J. Ind. Conv.* **2021**, *19*, 11–18.
29. Rai, A. Role of Perceived Customer Service, Quality, and Price Fairness on Attitude Formation: An Empirical Evidence from Nepal's Online Business Context. *J. Mgt.* **2022**, *5*, 01–10. [CrossRef]
30. Khandelwal, U.; Bajpai, N. Price fairness and its linear dependence on consumer attitude: A comparative study in metro and non metro city. *Euro J. Bus. Mgt.* **2012**, *4*, 94–102.
31. Sohaib, M.; Wang, Y.; Iqbal, K.; Han, H. Nature-based solutions, mental health, well-being, price fairness, attitude, loyalty, and evangelism for green brands in the hotel context. *Int. J. Hosp. Mgt.* **2022**, *101*, 103126. [CrossRef]
32. Theben, A.; Gerards, M.; Folkvord, F. The effect of packaging color and health claims on product attitude and buying intention. *Int. J. Env. Res. Pub Heal.* **2020**, *17*, 1991. [CrossRef] [PubMed]
33. Tudoran, A.; Olsen, S.; Dopico, D. The effect of health benefit information on consumers health value, attitudes and intentions. *Appe.* **2009**, *52*, 568–579. [CrossRef] [PubMed]
34. Jang, R.; Lee, W.; Moon, J. Relationship between Presentation, Attitude, and In-Flight Meal Food Healthiness: Moderating Role of Familiarity. *Food.* **2024**, *13*, 2111. [CrossRef]
35. Janssen, L.; Bogaert, E. Nutri-Score Vs. Nutrition Claim: The Effects of Incongruent Front-of-Pack Nutritional Information on Consumer Perceptions of Product Healthiness, Brand Attitude, and Purchase Intention. *J. Food Prod. Mkt.* **2023**, *29*, 69–81. [CrossRef]
36. Mohd Suki, N. Green product purchase intention: Impact of green brands, attitude, and knowledge. *Brit. Food J.* **2016**, *118*, 2893–2910. [CrossRef]
37. Han, H. Theory of green purchase behavior (TGPB): A new theory for sustainable consumption of green hotel and green restaurant products. *Bus. Str. Env.* **2020**, *29*, 2815–2828. [CrossRef]
38. Chen, Y.; Chang, T.; Li, H.; Chen, Y. The influence of green brand affect on green purchase intentions: The mediation effects of green brand associations and green brand attitude. *Int. J. Env. Res. Pub Heal.* **2020**, *17*, 4089. [CrossRef]
39. Zaremohzzabieh, Z.; Ismail, N.; Ahrari, S.; Samah, A. The effects of consumer attitude on green purchase intention: A meta-analytic path analysis. *J. Bus. Res.* **2021**, *132*, 732–743. [CrossRef]
40. Mountford-McAuley, R.; Prior, J.; Clavijo McCormick, A. Factors affecting propolis production. *J. Apic. Res.* **2023**, *62*, 162–170. [CrossRef]
41. Kasilingam, D.L. Understanding the attitude and intention to use smartphone chatbots for shopping. *Tech. Soc.* **2020**, *62*, 101280. [CrossRef]
42. Liesa-Orús, M.; Latorre-Cosculluela, C.; Sierra-Sánchez, V.; Vázquez-Toledo, S. Links between ease of use, perceived usefulness and attitudes towards technology in older people in university: A structural equation modelling approach. *Edu Inf. Tech.* **2023**, *28*, 2419–2436. [CrossRef] [PubMed]

43. Nuryakin, N.; Rakotoarizaka, N.; Musa, H. The effect of perceived usefulness and perceived easy to use on student satisfaction the mediating role of attitude to use online learning. *Asia Pac. Mgt Bus. Appl.* **2023**, *11*, 323–336. [CrossRef]

44. Madi, J.; Al Khasawneh, M.; Dandis, A. Visiting and revisiting destinations: Impact of augmented reality, content quality, perceived ease of use, perceived value and usefulness on E-WOM. *Int. J. Qual. Reli Mgt.* **2024**, *41*, 1550–1571. [CrossRef]

45. Iriani, S.; Andjarwati, A.L. Analysis of perceived usefulness, perceived ease of use, and perceived risk toward online shopping in the era of COVID-19 pandemic. *Sys Rev. Phar.* **2020**, *11*, 313–320.

46. Ahmad, W.; Kim, W.G.; Anwer, Z.; Zhuang, W. Schwartz personal values, theory of planned behavior and environmental consciousness: How tourists' visiting intentions towards eco-friendly destinations are shaped? *J. Bus. Res.* *110*, 228–236. [CrossRef]

47. Yarimoglu, E.; Gunay, T. The extended theory of planned behavior in Turkish customers' intentions to visit green hotels. *Bus. Str. Env.* **2020**, *29*, 1097–1108. [CrossRef]

48. Kim, S.; Jung, Y. Development of Semantic Differential Scales for Artificial Intelligence Agents. *Int. J. Soc. Rob.* **2023**, *15*, 1155–1167. [CrossRef]

49. Spatola, N.; Wudarczyk, O. Implicit attitudes towards robots predict explicit attitudes, semantic distance between robots and humans, anthropomorphism, and prosocial behavior: From attitudes to human–robot interaction. *Int. J. Soc. Rob.* **2021**, *13*, 1149–1159. [CrossRef]

50. Moon, J.; Lee, W.; Shim, J.; Hwang, J. Structural Relationship between Attributes of Technology Acceptance for Food Delivery Application System: Exploration for the Antecedents of Perceived Usefulness. *Systems* **2023**, *11*, 419. [CrossRef]

51. Hoyle, R. *Structural Equation Modeling: Concepts, Issues, and Applications*; Sage: Newcastle upon Tyne, UK, 1995.

52. Fornell, C.; Larcker, D.F. Evaluating structural equation models with unobservablevariables and measurement error. *J. Mkt Res.* **1981**, *18*, 39–50.

53. Hair, J.; Anderson, R.; Babin, B.; Black, W. *Multivariate Data Analysis: A Global Perspective (Vol. 7)*; Pearson: Upper Saddle River, NJ, USA, 2010.

foods

MDPI

Article

Bioactive Polyphenolic Compounds from Propolis of *Tetragonula carbonaria* in the Gibberagee Region, New South Wales, Australia

Dylan W. Ebner [1,2], Damon C. Woods [1,2] and Trong D. Tran [1,2,*]

[1] Centre for Bioinnovation, University of the Sunshine Coast, Sippy Downs, QLD 4556, Australia; debner@usc.edu.au (D.W.E.); dwoods1@usc.edu.au (D.C.W.)
[2] School of Science, Technology and Engineering, University of the Sunshine Coast, Sippy Downs, QLD 4556, Australia
* Correspondence: ttran1@usc.edu.au; Tel.: +61-7-5459-4579

Abstract: Stingless bee propolis has emerged globally as a new source of bioactive molecules that can advance human health. However, limited research has been conducted on Australian stingless bee propolis. This study investigated the chemical composition and biological activity of the propolis produced by the stingless bees *Tetragonula carbonaria* from Gibberagee, a distinct region of New South Wales state in Australia. Using bioassay-guided fractionation, twelve compounds were isolated, including six A-ring methylated flavonoids. Nine of these compounds demonstrated strong scavenging activity against 2,2-diphenyl-1-picrylhydrazyl radicals, with five exhibiting greater potency than vitamin C. Chemical structures of seven additional minor flavonoids were determined through an intensive MS/MS data analysis. In silico screening of these 19 compounds revealed that all, except for gallic acid, displayed a higher binding affinity to α-glucosidase than the antidiabetic drug, voglibose. This study showed that the Gibberagee stingless bee propolis is a promising source for nutraceutical and cosmeceutical applications owing to its strong antioxidant and antidiabetic properties. The unique profile of A-ring C-methylated flavonoids potentially provides valuable insights into its botanical origin.

Keywords: *Tetragonula carbonaria*; stingless bee propolis; flavonoids; polyphenolics; antioxidant; antidiabetic

check for updates

Academic Editors: Liming Wu and Qiangqiang Li

Received: 9 February 2025
Revised: 4 March 2025
Accepted: 10 March 2025
Published: 12 March 2025

Citation: Ebner, D.W.; Woods, D.C.; Tran, T.D. Bioactive Polyphenolic Compounds from Propolis of *Tetragonula carbonaria* in the Gibberagee Region, New South Wales, Australia. *Foods* **2025**, *14*, 965. https://doi.org/10.3390/foods14060965

1. Introduction

Stingless bees are a large group of social bees [1]. To date, there are more than 500 known stingless bee species distributing in the tropical and subtropical regions including the Neotropical region of South America (approximately 391 species), the Indo-Malayan region of Asia (approximately 60 species), the Paleotropical region of Africa (approximately 50 species), and the Australasia region of Oceania (approximately 11 species) [1]. Unlike honeybees, which build their nests primarily or even solely out of wax [2], most stingless bees use propolis, which is a mixture of plant resin and beeswax, for nest construction [3]. Compared to other bee products, such as honey, pollen, and beeswax, propolis contains the highest concentration of specialised plant metabolites and has valuable pharmacological activities [4]. The stingless bee propolis has been used in traditional medicines in Mexico, Brazil, Argentina, India, and Vietnam as a remedy for improving health and treating diseases [5]. Modern research has confirmed the biological activities of stingless bee propolis, including antioxidant, antimicrobial, anti-inflammatory and anticancer activities, which have a strong linkage with the chemical compositions of the original plant resin [6]. The chemical composition of stingless bee propolis consists of mainly polyphenolic and terpenoid compounds, whose ratio varies depending on the propolis type [7]. Possessing potent and wide-spectrum biological activities and diverse chemical composition, the propolis of stingless bees is considered a good resource for functional food and nutraceutical ingredients and potential novel pharmaceutical candidates [6,7].

Australia has more than 1700 native bee species, of which 11 are stingless and belong to the genera *Tetragonula* and *Austroplebeia* [8]. The domestication of *Tetragonula carbonaria* (*T. carbonaria*) colonies began in the 1980s [9], and previous propolis research has only focused on samples of this species collected in Queensland, Australia [5]. Massaro and her colleagues reported that propolis from *T. carbonaria* had an anti-inflammatory property with a distinct chemical profile compared to propolis from honeybee *Apis mellifera* [10]. This propolis extract was also found to relax porcine coronary arteries in an endothelial-independent manner [11]. Chemical investigation of the *T. carbonaria* propolis harvested in South East Queensland resulted in the isolation of six flavanones, including (2S)-cryptostrobin, (2S)-stroboponin, (2S)-cryptostrobin-7-methyl ether, (2S)-desmethoxymatteucinol, (2S)-pinostrobin, and (2S)-pinocembrin [12]. These compounds showed antimicrobial activity against the Gram-positive *Staphylococcus aureus*, with MIC values ranging from 6.9 to 182.2 µg/mL [12]. Two novel phloroglucinols were found in the *T. carbonaria* propolis [13]. Although their biological activity has not been reported, the identification of these two compounds demonstrated the potential of finding novel molecules from Australian native stingless bee propolis, which originates from unique botanical sources. More recently, a potent antioxidant meroterpene, tomentosenol A, was identified from the *T. carbonaria* propolis [14]. This compound showed significant antifibrotic potential via the inhibition of transforming growth factor-β1 (TGF-β1)-stimulated, NFF-myofibroblast differentiation and soluble collagen production [14].

In recent years, polyphenolic compounds have attracted significant interest as possible therapeutic remedies and disease prevention against oxidative stresses linked with many chronic diseases such as diabetes mellitus, cancer, neural degradation, and cardiovascular diseases [15,16]. With diabetes mellitus being one of the world's leading global health issues, leading to an increase of 80% risk of mortality for those with the disease, the need for more diverse disease prevention and management has become increasingly urgent [17]. One of the common treatments for diabetes mellitus is to retard the absorption of glucose through the inhibition of the enzyme α-glucosidase. While common therapeutic drugs (including acarbose, miglitol, and voglibose) that strongly inhibit this enzyme are readily available, these drugs often come with side effects, including diarrhoea, abdominal distention, and nausea [18]. Polyphenols have also been shown in the previous literature to potentially inhibit α-glucosidase, suggesting that incorporating additional polyphenols into the diet may reduce the need for these drugs and the progression and risk of diabetes mellitus [19,20]. With the aim to investigate the bioactive composition of the Australian native stingless bee propolis, this paper reports the potential antioxidant and antidiabetic compounds of the *T. carbonaria* propolis collected in Gibberagee, the Northeast region of New South Wales, Australia. Bioassay-guided fractionation led to the isolation of 12 phenolic and flavonoid compounds. Molecular networking analysis further revealed the presence of other flavonoid compounds in this propolis extract. The antidiabetic activities of these compounds were then assessed via an in silico assay.

2. Materials and Methods

2.1. Solvents and Reagents

Solvents used for extraction (ethanol—EtOH), HPLC (acetonitrile—MeCN), and LC-MS (MeCN and water—H_2O) analyses were purchased from Merck (Melbourne, Australia).

Ultra-pure water used for HPLC analysis was from an in-house Milli-Q system. The reagents, 2,2-diphenyl-1-picrylhydrazyl (DPPH) and L-ascorbic acid; NMR solvents, including deuterated chloroform (CHCl$_3$-*d*) and deuterated methanol (MeOH-*d*$_4$); and formic acid were purchased from Sigma-Aldrich (Melbourne, Australia).

2.2. Sample Collection and Extraction

Raw propolis from stingless bees *T. carbonaria* was harvested in June 2022 and stored in darkness at 4 °C. The raw propolis was then frozen at −20 °C, and the sample was powdered by grinding manually. Fine propolis powder (0.5 g × 10) was mixed with 5 mL of 70% (*v*/*v*) ethanol solution, heated at 65 °C for 30 min and then sonicated in an ultrasonic bath for 5 min. The sample was placed in ice for 10 min before being centrifuged at 3600 rpm at 4 °C for 10 min. The supernatant was dried under a vacuum using a GeneVac EZ-2 evaporator (Genevac, Ipswich, UK) to obtain dry propolis extract. Dry extracts were then combined for compound isolation.

2.3. Isolation and Purification

The extract (392.9 mg) was fractionated using a C$_{18}$ Synergi Fusion HPLC column (4 µm, 100 × 21.2 mm) at a flow rate of 10 mL/min. The mobile phase consisted of H$_2$O (solvent A) and MeCN (solvent B), running for 50 min with a linear gradient starting at 5% B for 10 min and increasing to 100% B for 30 min and then running isocratic for another 10 min to give six fractions, A–F (8.0 min for each fraction A–E, and 10 min for fraction F), which were collected. Fractions B–E showed free radical scavenging activity. Fraction B was purified on a Synergi Fusion HPLC column (4 µm, 100 × 21.2 mm) at a flow rate of 10 mL/min with a linear gradient from 5% B to 25% B for 30 min to yield compound **1** (2.2 mg, t_R = 15 min, 0.56%). Fraction C was loaded on the same Synergi Fusion HPLC column at a flow rate of 10 mL/min with a linear gradient from 20% B to 50% B for 30 min to obtain compounds **2** (3.2 mg, t_R = 18 min, 0.81%) and **3** (1.2 mg, t_R = 22 min, 0.31%). Fraction D was purified on the Synergi Fusion HPLC column at a flow rate of 10 mL/min with a linear gradient from 25% B to 60% B for 30 min to yield compounds **4** (4.2 mg, t_R = 20 min, 1.1%) and **5** (1.6 mg, t_R = 25 min, 0.41%). Fraction E chromatogram was obtained on the Inertsil Diol column (5 µm, 250 × 20 mm) at a flow rate of 10 mL/min using 80% hexane/20% isopropanol (solvent A) and 100% hexane (solvent B) as a mobile phase. The HPLC purification was run for 60 min with a linear gradient starting from 0% A to 25% A for 10 min, increasing to 65% A for 30 min and then to 80% A for the next 10 min and rising to 100% in the last 10 min to yield compound **6** (1.1 mg, t_R = 31 min, 0.28%), compound **7** (4.1 mg, t_R = 38 min, 1.0%), compound **8** (1.9 mg, t_R = 45 min, 0.48%), compound **9** (0.8 mg, t_R = 46 min, 0.20%), compound **10** (3.7 mg, t_R = 47 min, 0.94%), compound **11** (0.7 mg, t_R = 56 min, 0.18%), and compound **12** (0.8 mg, t_R = 59 min, 0.20%).

2.4. Evaluation of Antioxidant Activity Using DPPH Free Radical Scavenging Assay

The DPPH free radical scavenging activity of the propolis extracts at different concentrations was evaluated using the DPPH assay as described previously [21]. Briefly, the DPPH solution was prepared on the day of measuring at a concentration of 100 µM in MeOH. The propolis extracts (200 µL) at different concentrations were added to 600 µL of DPPH solution in Eppendorf tubes. The mixtures were kept in the dark at room temperature for 20 min before being plated to a 96-well plate (200 µL/well) and measured at 518 nm using a Perkin Elmer Enspire microplate reader (Waltham, MA, USA). All evaluations were performed in triplicate. Gallic acid and MeOH were used as positive and negative controls.

The percentage of inhibition of the DPPH radical for each sample was normalised and calculated using the following formula:

$$\% \ Inhibition = \left[1 - \frac{(A_S - A_P)}{(A_B - A_P)} \right] \times 100$$

where A_S is the absorbance of the sample, A_P is the absorbance of the positive control, and A_B is the absorbance of the blank sample (negative control).

An absolute IC_{50} curve for each compound was generated using GraphPad Prism 10 (GraphPad, Boston, MA, USA) with a 95% confidence interval. The IC_{50} values were determined as the concentration required to inhibit 50% of DPPH free radicals and reported as mean $\pm$ standard deviation.

2.5. NMR Analysis

The NMR spectra were acquired on a Bruker Ascend 400 spectrometer (Billerica, MA, USA) equipped with a 5 mm room temperature probe operating at 400 MHz for ^{1}H NMR and 100 MHz for ^{13}C NMR. All experiments were acquired in automation (temperature equilibration to 298 K, optimisation of lock parameters, gradient shimming, and setting of receiver gain). Compounds **1–12** were dissolved in CHCl$_3$-*d* or MeOH-*d$_4$*. The ^{1}H and ^{13}C spectra were referenced to the residual deuterated solvent peaks at δ_H 7.26 and δ_C 77.0 (CHCl$_3$-*d*) and δ_H 3.31 and δ_C 49.0 (MeOH-*d$_4$*).

2.6. LC-QTOF MS Analysis

All LC-MS analyses were performed using an analytical scale Agilent 1290 uHPLC system combined with an Agilent 6546 QTOF mass spectrometer (Santa Clara, CA, USA). Separations were performed at 35 °C on a Zorbax Eclipse Plus C$_{18}$ column (50 × 2.1 mm, 1.8 μm particle size, 95 Å pore size) with a flow rate of 0.4 mL/min. The mobile phase consisted of H$_2$O (solvent A), and MeCN (solvent B), both acidified with 0.1% formic acid. The samples were separated using a 15 min programme, which started at 2% B for 0.5 min, increased to 100% B for 9 min, kept at this level for the next 3 min, reduced to 2% B for 1 min, and re-equilibrated for 1.5 min. The injection volume was 2 μL. The mass spectrometer is equipped with an ESI source. Mass spectra were acquired in both positive and negative ionisation modes using gas temperature of 250 °C, a gas flow of 5 L/min, a capillary voltage of 4000 V, a nebuliser pressure of 30 PSI, a sheath gas heater of 400 °C, a sheath gas flow of 12 L/min, and a nozzle voltage of 1000 V. Chromatographic separation and mass spectrometry were controlled using the Mass Hunter software (B.10.00, Agilent Technologies, Santa Clara, CA, USA).

2.7. Molecular Networking Analysis

The MS and MS/MS data acquired by Agilent uHPLC-QTOF MS were converted to mzML format using MSConvert (version: 3.0.23244-bc8a3ad) as part of the ProteoWizard suite and uploaded to the MassIVE MS data repository (https://www.nature.com/articles/nbt.3597, accessed on 15 August 2024). The Global Natural Product Social Molecular Network (GNPS) was created using the online workflow (https://ccms-ucsd.github.io/GNPSDocumentation/, accessed on 15 August 2024) on the GNPS website (http://gnps.ucsd.edu, accessed on 15 August 2024) [22]. The data were filtered by removing all MS/MS fragment ions within +/− 17 Da of the precursor m/z. MS/MS spectra were window-filtered by choosing only the top 6 fragment ions in the +/− 50 Da window throughout the spectrum. The precursor ion mass tolerance was set to 2.0 Da, and an MS/MS fragment ion tolerance of 0.5 Da was set. A network was created where edges were filtered to have a cosine score above 0.7 and more than four matched peaks. Further, edges between two nodes were kept in the network if and only if each of the nodes appeared in each other's respective top ten most similar nodes. Finally, the maximum size of a molecular family was set to 100, and the lowest-scoring edges were removed from molecular families until the molecular family size was below this threshold. The spectra in the network were then searched against GNPS spectral libraries. The library spectra were filtered in the same manner as the input data. All matches kept between network spectra and library spectra were required to have a score above 0.7 and at least four matched peaks.

The GNPS output was then visualised using Cytoscape (version 3.10.2). The resulting retention times and precursor mass (+) m/z and (−) m/z from SIRIUS and GNPS outputs were matched to identify previously identified compounds through NMR analysis. Unknown compounds were then further identified using the differences in precursor mass and their MS/MS relationship with the neighbouring compounds. The Cytoscape data were then adapted for use in Biorender (version 2025). The GNPS original data can be accessed through the Supplementary Materials.

The SIRIUS software (version 5.8.6) was downloaded from the Lehrstuhl Bioinformatik Jena website (https://bio.informatik.uni-jena.de/software/sirius/, accessed on 1 December 2024). The MS and MS/MS data acquired by uHPLC-QTOF MS (6546 Agilent) was converted to mzML format using MSConvert (version: 3.0.23244-bc8a3ad) as part of the ProteoWizard suite. The mzML files contained the m/z of each compound and its relative intensity (%) extracted directly from the uHPLC-QTOF MS (6546 Agilent) using the Agilent MassHunter Workstation Software version B.08.00. To compute the molecular formulas, instrument type was set as Q-TOF; mass accuracy was set as 10 ppm; possible ionisation for positive mode was selected as [M+H]+, [M+K]+, and [M+Na]+; and possible ionisation for negative mode was selected as [M−H]−, [M−K]− and [M−Na]−. C, H, and O were selected for element searches, and the number of candidates was set to 10. Database formulas used were CHEBI, COCONUT, GNPS, KEGG, KEGG Mine, KnaPSnaCK, KnaP-SnaCK, Maconda, Natural Products, PlantsCYC, PUBCHEM, and PUBMED. Structure elucidation by CSI: FingerID was set to search using the same adducts as SIRIUS and the same database sets. Canopus Class Prediction was also enabled.

2.8. In Silico Screening of Compounds Against α-Glucosidase

The Protein Data Bank structure of α-glucosidase from *Saccharomyces cerevisiae* (PDB code: 2ZQ0) was downloaded from the Research Collaboratory for Structural Bioinformatics (RCSB) Protein Data Bank (http://www.rcsb.org, accessed on 15 December 2024). The protein structure was analysed using Discovery Studio Visualiser 4.5, and its dimer was removed. Using galaxy.org.au, water molecules, heteroatoms, and ligands were removed, and polar hydrogen atoms were added (pH 7.4) to the structures as described by the official tutorials. The proteins' binding and active site residues were determined as previously described by Uddin et al. [23] and then the binding site sphere was defined accordingly. The dimension of the sphere was 20 Å, the sphere's centre (x, y, z) was (27.776172, 56.165741, 35.362259), and the exhaustiveness value was adjusted to 24. Binding affinity was calculated using galaxy.org.au [24]. The top 9 binding poses were opted for prediction, and results were analysed using Discovery Studio Visualiser 4.5. The docking study was validated by redocking and superimposing known α-glucosidase inhibitors (acarbose and vogibose) with the extracted protein from the crystal structure.

3. Results and Discussion

3.1. Antioxidant Activity of the Gibberagee Propolis and the Pure Polyphenols

The propolis extract exhibited a scavenging property against the free radical DPPH by 80% at a concentration of 100 μg/mL and displayed an IC_{50} value of 24.5 μg/mL (Table 1).

Bioassay-guided fractionation was employed to isolate one phenolic acid, gallic acid (**1**) [25], and eleven flavonoids, including catechin (**2**) [26,27], epicatechin (**3**) [26,27], myricetin (**4**) [28], eriodictyol (**5**) [29], pinocembrin (**6**) [30], cryptostrobin (**7**) [12], myrigalone H (**8**) [31], strobopinin (**9**) [12], angophorol (**10**) [32], 8-methylsakuranetin (**11**) [32], and sakuranetin (**12**) [33]. Their chemical structures are depicted in Figure 1 and were verified by comparing their NMR and MS data with previous data reported in the literature.

Table 1. Antioxidant activity of isolated compounds in *T. carbonaria* propolis from Gibberagee, Australia.

Extract/Compounds	% Inhibition at 100 µg/mL	IC$_{50}$ (µg/mL)	IC$_{50}$ (µM)
Propolis extract	80 ± 2	24.5 ± 0.3	-
Gallic acid (**1**)	100 ± 1	0.9 ± 0.1	5.3 ± 0.6
Catechin (**2**)	100 ± 0	2.5 ± 0.2	8.6 ± 0.7
Epicatechin (**3**)	99 ± 1	2.3 ± 0.1	7.9 ± 0.3
Myricetin (**4**)	100 ± 1	1.8 ± 0.1	5.7 ± 0.3
Eriodictyol (**5**)	100 ± 0	3.4 ± 0.2	11.8 ± 0.7
Pinocembrin (**6**)	38 ± 3	N.D [a]	N.D [a]
Cryptostrobin (**7**)	41 ± 5	N.D [a]	N.D [a]
Myrigalone H (**8**)	80 ± 2	20.0 ± 0.5	69.9 ± 1.7
Strobopinin (**9**)	40 ± 1	N.D [a]	N.D [a]
Angophorol (**10**)	63 ± 4	58.4 ± 0.8	186 ± 2.5
8-Methylsakuranetin (**11**)	56 ± 4	85.2 ± 1.1	284 ± 3.7
Sakuranetin (**12**)	50 ± 1	100 ± 0.1	350 ± 0.3
Ascorbic acid (Vitamin C)	100 ± 0	5.0 ± 0.1	28.4 ± 0.6

N.D [a]: Not determined due to lower 50% inhibition against DPPH at 100 µg/mL, which was the maximum concentration tested.

The DPPH radical scavenging assay was used to evaluate the antioxidant activities of all isolated compounds, as shown in Table 1. Nine of the twelve compounds exhibited greater than 50% inhibition of DPPH radicals at 100 µg/mL, and five of them showed more potent antioxidant activity than L-ascorbic acid (vitamin C) (Table 1). Gallic acid (**1**) demonstrated the most potent antioxidant activity with an IC$_{50}$ value of 0.93 µg/mL. Among the 11 flavonoids tested, the flavonol myricetin (**4**) showed the strongest DPPH radical scavenging activity (IC$_{50}$ of 1.84 µg/mL) and was followed by two flavanols with almost similar antioxidant potency, epicatechin and catechin (IC$_{50}$ of 2.28 and 2.49 µg/mL, respectively). A direct comparison with the positive control, vitamin C, demonstrates that compounds **1–5** exhibit more potent DPPH radical scavenging activity, with IC$_{50}$ values ranging from 5.3 to 11.8 µM compared to 28.4 µM for vitamin C. The data of isolated flavonoids indicated that the number and position of the hydroxyl groups in the rings were crucial to maintaining antioxidant activity. The results were consistent with previous studies in which the presence of a catechol group in ring B (compounds **2–5**) and a 3-hydroxyl group in a heterocyclic ring C increased radical scavenging activity, and a 2,3-double bond conjugated with the 4-oxo group in the ring C strengthened the antioxidant activity [34]. No significant difference in the antioxidant activity was observed when a methyl group was attached to ring A (compounds **7** and **9–11**). The double inhibition of DPPH at 100 µg/mL between a chalcone, myrigalone H (**8**) (80%), and a flavanone, pinocembrin (38%), indicated that opening the ring C facilitated the scavenging potency. Additionally, the release of a resorcinol moiety in ring A potentially leads to the enhancement of the antioxidant activity.

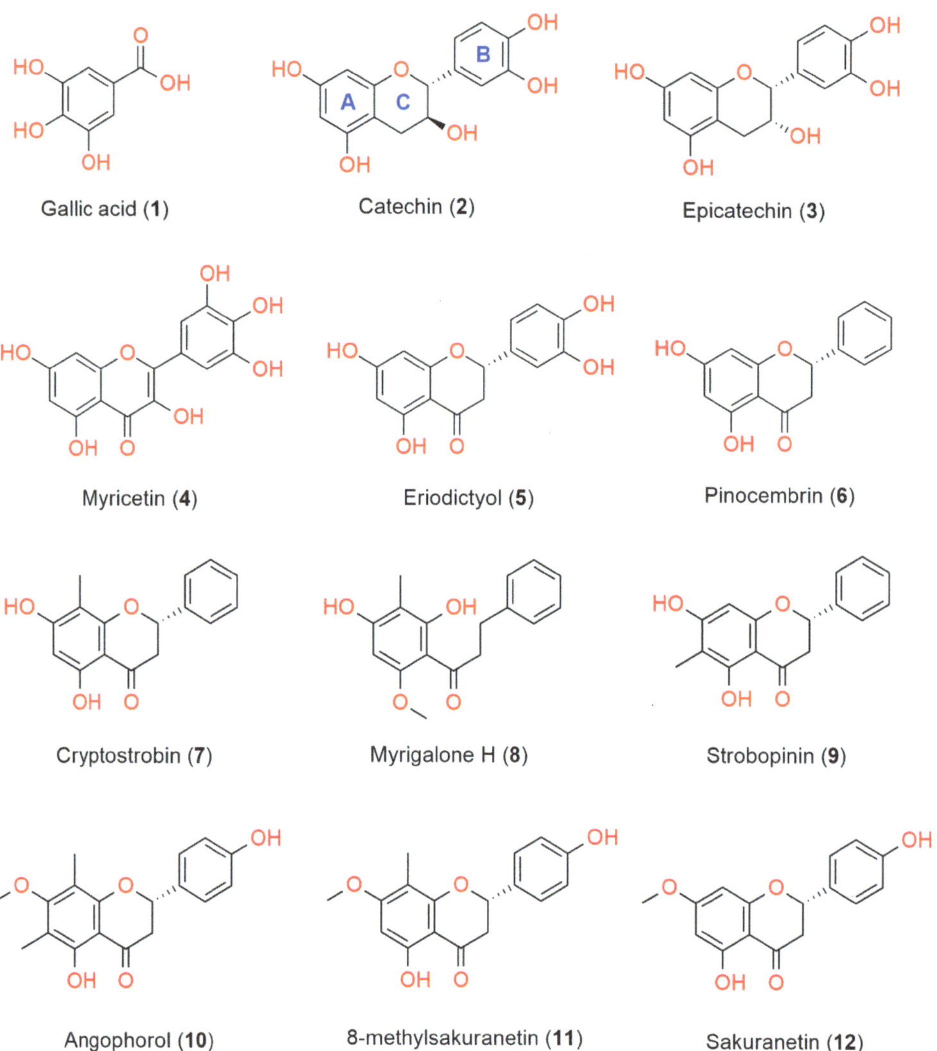

Figure 1. Chemical structures of compounds isolated from the propolis of stingless bees *Tetragonula carbonaria* from Gibberagee, Australia. (A, B and C in blue are labelled for the flavonoid ring system).

3.2. Mining Polyphenolic Compounds Through Molecular Networking Analysis

The compound structures depicted in orange in Figure 2 were determined from the elucidation of their MS/MS data and the comparison of their MS/MS data with those of isolated compounds (**1–12**). Their structures were then further confirmed by searching the SIRIUS compound database (Table S1 in Supplementary Materials).

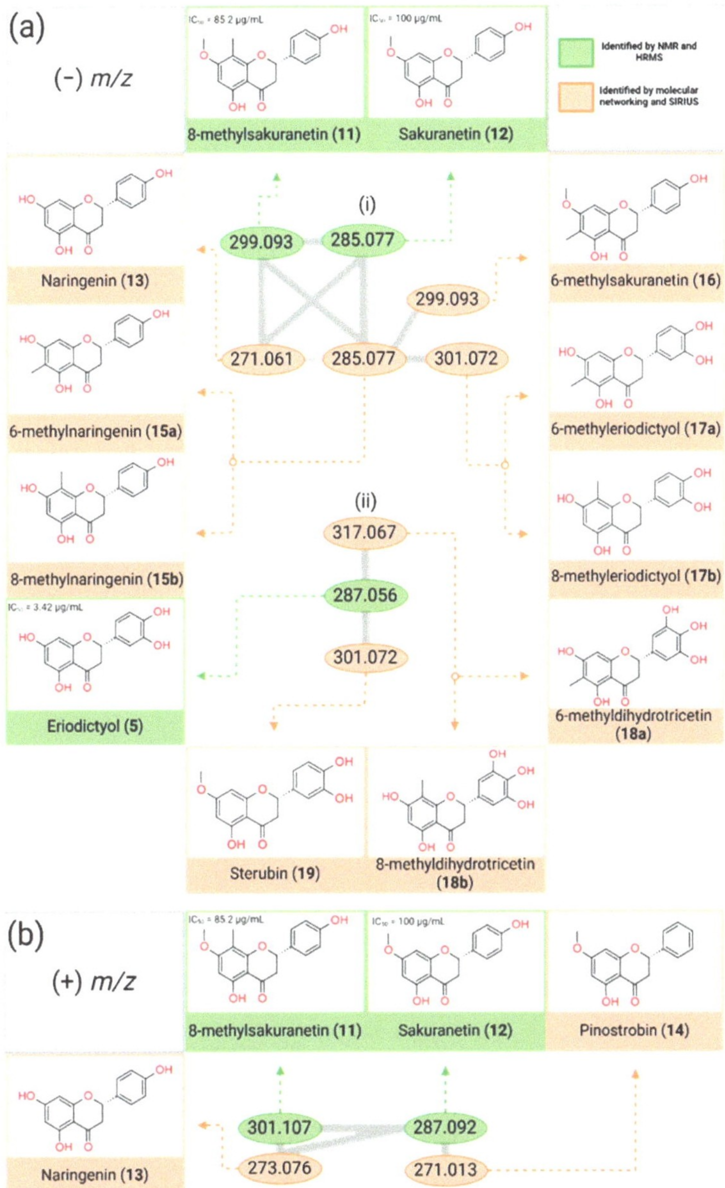

Figure 2. Molecular networking analysis of *T. carbonaria* propolis collected from Gibberagee, Australia, from the negative mode (**a**) and the positive mode (**b**); original network created using GNPS and visualised in Cytoscape, and the network was then recreated in BioRender.com. Green: isolated compounds; orange: tentative compounds. Isolated compounds (Green) have their IC_{50} values that were reported from the DPPH assays performed in this study.

Using the negative MS/MS data, GNPS was able to network two clusters of flavonoids containing six (Figure 2a(i)) and three (Figure 2a(ii)) nodes in the negative mode (Figure 2a). Figure 2a(i) further confirmed the presence of 8-methylsakranetin (**11**) ((−) *m/z* 299.093) and sakuranetin (**12**) ((−) *m/z* 285.077). Four additional flavanones that were not isolated using HPLC were assigned from this molecular networking analysis. These compounds included naringenin (**13**) ((−) m/z 271.061), 6-methylnaringenin (**15a**) or 8-methylnaringenin (**15b**) ((−) m/z 285.077), 6-methylsakuranetin (**16**) ((−) m/z 299.093), and 6-methyleriodicyol (**17a**) or 8-methyleriodicyol (**17b**) ((−) m/z 301.072). The negative cluster (Figure 2a(ii)) elucidated an additional two compounds, sterubin (**19**) ((−) m/z 301.072) and 6-methyldihydrotricetin (**18a**) or 8-methyldihydrotricetin (**18b**) ((−) m/z 317.067).

Using the positive MS/MS data, GNPS created a single cluster of flavonoids (Figure 2b) containing four nodes that were then annotated. Figure 2b further confirmed the presence of compounds 8-methylsakranetin (**11**) ((+) m/z 301.107) and sakuranetin (**12**) ((+) m/z 287.092). From this positive cluster, two additional compounds that were not isolated using HPLC were identified, including naringenin (**13**) ((+) m/z 273.076) and pinostrobin (**14**) ((+) m/z 271.013).

Based on the compounds determined, it is evident that *T. carbonaria* bees in the Gibberagee region have an affinity for flavonoids that have methyl and methoxy groups on the A-ring. As Australia is considered a megadiverse country, the flora surrounding these bees is likely very different from that in overseas studies. As a result, the native bees of the Gibberagee region collect different plant materials to make their propolis. This distinct difference in the flavonoids may also be attributed to the difference in the bee species enzymes; as they make the propolis, they mix their saliva with the plant material, likely introducing chemical reactions between the saliva and plant material, providing the bees with distinct differences in their propolis composition. *Melichrus gibberagee* and *Eucalyptus punctata* are two Australian endemic plant species that dominate the Gibberagee region [35]. These plants may provide a rich source of flavonoid compounds for native stingless bees to forage, warranting further studies to identify the botanical source of the *T. carbonaria* propolis in Gibberagee.

3.3. In Silico Assay

α-glucosidase is one of the most important enzymes in carbohydrate digestion, as this enzyme is primarily responsible for the breakdown of complex carbohydrates, such as oligosaccharides, into monosaccharides. The inhibition of this enzyme leads to the delayed release of glucose into the bloodstream. Drugs such as acarbose and voglibose or nutraceuticals that offer similar effects can be instrumental in lowering postprandial blood sugar levels in those with diabetes mellitus [36].

To understand if compounds identified from Gibberagee propolis have a potential antidiabetic property, they were subjected to an in silico assay against the α-glucosidase enzyme (PDB ID: 2ZQ0). The docked complex is considered the best-docked if it exerts the most negative binding affinity energy value, reflecting a strong protein–ligand interaction and thereby potentially blocking the active site of the enzyme. The binding affinity values and detailed interactions at the binding site of the 19 compounds, along with two positive controls (acarbose and voglibose) towards α-glucosidase, are available in Table S2 (Supplementary Materials). From the docking results, it was found that the positive control, acarbose, exhibited the strongest binding affinity (−10.049 kcal/mol), and, among the identified compounds in this study, 8-methyldihydrotricetin (**18b**) exhibited the strongest binding affinity (−9.578 kcal/mol). The 19 compounds in this study showed affinity energy ranging from −9.578 kcal/mol to −5.981 kcal/mol. All compounds docked in this study showed higher binding affinity towards the enzyme than voglibose, except for gallic acid (**1**) (−7.264 versus −5.981 kcal/mol).

The docking study indicated that acarbose and voglibose largely depend on hydrogen–hydrogen bonding within the active site of the enzyme (Figure 3a). However, the polyphenolics were found to interact with the active site of the enzyme utilising ring B (Figure 3b), forming electrostatic bonds (π-anion) with the amino acid GLU^{439} and/or GLU^{532}. All the polyphenolic compounds in this study formed either π–alkyl or π–π bonds between ring B of the polyphenol and the amino acid VAL[471] and PHE[536] (Table S2). These two bonds appear to be the foundation interactions that allow polyphenols to inhibit α-glucosidase. When the compounds have a hydroxy in ring B, this hydroxy creates a hydrogen bond with GLU[391], enhancing the binding with α-glucosidase. The C-methylation on ring A showed some effects on the binding affinity. C-methylation at position C-8 exhibited a stronger interaction between the polyphenol and the enzyme when compared with the C-methylation at position C-6. This can be seen when comparing compounds **11** (−8.857 kcal/mol) versus **16** (−8.449 kcal/mol), **15b** (−8.275 kcal/mol) versus **15a** (−8.190 kcal/mol), **17b** (−9.224 kcal/mol) versus **17a** (−9.216 kcal/mol), and **18b** (−9.578 kcal/mol) versus **18a** (−9.538 kcal/mol).

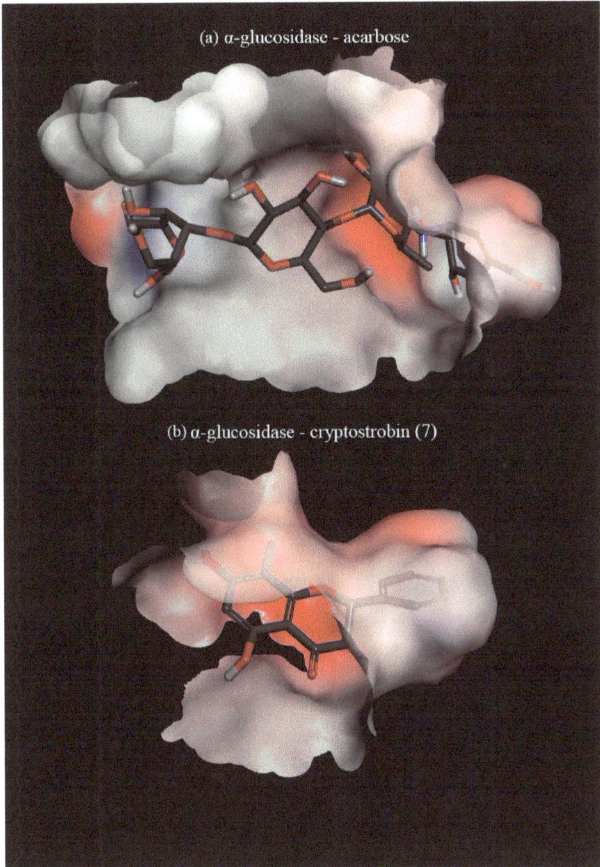

Figure 3. (**a**) Molecular docking of the protein α-glucosidase (2ZQ0) with acarbose (control); (**b**) molecular docking of the protein α-glucosidase (2ZQ0) with cryptostrobin (**7**). Visualised using Discovery Studio.

The relationship between ring A substitution of flavones and α-glucosidase inhibition was studied extensively in vivo by Gao and his colleagues [37]. It was observed that without any ring B substitutions, ring A hydroxylations at positions C-5, C-6, and C-7 were crucial for α-glucosidase inhibition and that most substitutions at position C-8 reduced the effects of the compound's ability to inhibit α-glucosidase [37]. However, potential confounding factors between both ring A and ring B hydroxylation, methoxylation, and C-methylation substitutions have not been identified. In our in silico study, an increase in binding affinity with the addition of the C-methylation substitution was observed, as evidenced by the comparison of binding affinity between compounds **11** (−8.857 kcal/mol) and **12** (−8.202 kcal/mol) and between compounds **7** (−8.908 kcal/mol) and **6** (−8.579 kcal/mol).

A previous study also suggested that a pair of hydroxy groups on ring B in positions C-3′ and C-4′ increase the activity of flavonoids in inhibiting α-glucosidase [38]. During the docking of the compounds that were identified, a general trend of the more hydroxy groups on the B ring, the greater the binding activity, was also found, as evidenced by the binding affinity of compounds **17a**, **17b**, **18a**, **18b**, and **19**. This can be further confirmed when comparing the binding affinity of compounds **12** (−8.202 kcal/mol) to **19** (−8.734 kcal/mol), whose difference is the addition of another hydroxy group on ring B at position C-5′. As seen from previous studies on the antioxidant activity of flavonoids, an increase in the hydroxy groups on ring B leads to an increase in free radical scavenging [39,40].

From the results of this molecular docking study, there appears to be a positive relationship between the number of hydroxy groups on ring B, the polyphenols, and the binding affinity of α-glucosidase, further linking the relationship between antioxidant activity and α-glucosidase activity. These findings would encourage further in vitro and in vivo screening to verify the antidiabetic properties of this propolis type and the unique methylated flavonoids present within it.

3.4. Potential Bioactivity of Compounds Identified from the Gibberagee Propolis

A literature search on the compounds identified from the Gibberagee propolis demonstrated that many of the compounds exhibited a wide variety of bioactivities, including anti-inflammatory, antioxidant, anticancer, antimicrobial, neuroprotective, antiallergic, anti-angiogenic, anticancer, antiviral, cardioprotective, and antidiabetic properties (Table 2).

Table 2. Known bioactivities of compounds identified in *T. carbonaria* propolis from Gibberagee, Australia, as reported in the literature.

Compound	Known Bioactivities
Gallic Acid (**1**)	Antiallergic [41], Anti-angiogenic [42], Anticancer [43], Antidiabetic [44,45], Anti-inflammatory [46], Antimicrobial [43,44], Antioxidant [44], Antiviral [44], Neuroprotective [44]
Catechin (**2**)	Cardiovascular Protective [47], Anticancer [48], Antidiabetic [49], Anti-inflammatory [50], Antimicrobial [51], Antioxidant [47], Neuroprotective [47]
Epicatechin (**3**)	Anticancer [47], Antidiabetic [49,52], Antioxidant [47], Cardiovascular protective [47], Neuroprotective [52]
Myricetin (**4**)	Anticancer [53], Antidiabetic [53], Antihypertensive [53], Antimicrobial [53], Antioxidant [53], Immunomodulatory [53], Neuroprotective [53]
Eriodictyol (**5**)	Anticancer [54], Antidiabetic [54], Anti-inflammatory [54], Antioxidant [54], Cardioprotective [54], Hepatoprotective [54], Neuroprotective [54]
Pinocembrin (**6**)	Anticancer [55], Antifibrotic [55], Anti-inflammatory [55], Antimicrobial [55], Antioxidant [55], Cardiovascular protective [55], Neuroprotective [55]
Cryptostrobin (**7**)	Antibacterial [12,56], Antidiabetic [57], Antihypertensive [56]
Myrigalone H (**8**)	Antibacterial [58]
Strobopinin (**9**)	Anti-inflammatory [59], Antimicrobial [59], Antioxidant [59], Antiparasitic [59], Neuroprotective [60]

Table 2. *Cont.*

Compound	Known Bioactivities
Angophorol (**10**)	Anticancer [61]
Sakuranetin (**12**)	Antiallergic [62], Anticancer [62], Anti-inflammatory [62], Antimicrobial [62], Antimutagenic [62], Antioxidant [62], Antiparasitic [62], Antiviral [62]
Naringenin (**13**)	Anticancer [63], Antidiabetic [63], Antimicrobial [63], Antidiabetic [63], Antioxidant [63], Cardiovascular protective [63], Gastroprotective [63], Immunomodulatory [63], Neuroprotective [63]
Pinostrobin (**14**)	Antibacterial [64], Anticancer [64,65], Antidiabetic [65], Anti-inflammatory [65], Antioxidant [64], Antiviral [64]
Sterubin (**19**)	Anti-inflammatory [66], Antioxidant [66], Neuroprotective [66]

The antidiabetic effects of eight polyphenolics (compounds **1–5**, **7**, **13**, and **14**) were previously determined through in vitro assays. Although gallic acid (**1**) did not show a strong interaction with α-glucosidase (-5.981 kcal/mol) in the in silico assay compared to other compounds, previous research reported that this compound significantly improved both the antioxidant status and glucose homeostasis of diabetic mice [45]. Compounds **11**, **15a**, **15b**, **16**, **17a**, **17b**, **18a**, and **18b** have been sparsely studied in the literature regarding their potential bioactivities, and no antidiabetic activity has been previously reported for these compounds. The unique combination of polyphenolics with known and unknown antidiabetic properties in the Gibberagee propolis warrants further investigations of its activity through in vitro and in vivo assays.

4. Conclusions

The combination of the bioassay-guided fractionation and the GNPS molecular networking analysis led to the identification of several polyphenol subclasses, including flavanones, flavanols, flavanols, and chalcones, with a unique mixture of C-methylated flavonoids on the A-ring for the Gibberagee propolis. Some of these compounds have not been previously investigated for their bioactivities. Five of the identified compounds exhibited 2- to 5-fold greater free radical scavenging activity than vitamin C, while eleven compounds demonstrated stronger α-glucosidase inhibition. The identification of the antioxidant and α-glucosidase inhibitory properties of polyphenolic compounds in this propolis highlights its potential for the development of nutraceutical and cosmeceutical applications, inspiring new avenues of research and product development for this unique propolis. Further in vitro and in vivo studies will be required to validate its therapeutic properties. Additionally, the identification of the botanical sources for this propolis by examining unique flora in the Gibberagee region, such as *Melichrus gibberagee* and *Eucalyptus punctata*, will contribute to the conservation of Australian native plant species.

Supplementary Materials: The following supporting information can be downloaded at https://www.mdpi.com/article/10.3390/foods14060965/s1, Table S1: Description of compounds identified from Gibberagee stingless bee propolis using a combination of GNPS and SIRIUS from negative and positive mode MS/MS data; Table S2: Binding affinity and binding interaction of compounds **1–19** to α-glucosidase (2ZQ0); Figure S1: Global Natural Products Social Molecular Networking non-clustered compounds visualised in Cytoscape (version 3.10.3); Figure S2: Non-covalent interactions of sakuranetin (**12**) and acarbose (control) with α-glucosidase (2ZQ0); Figure S3: Comparisons between the MS^1 spectra of some identified compounds and those in the SIRUS databases.

Author Contributions: T.D.T.: Conceptualisation, visualisation, project administration, funding acquisition, supervision; D.W.E.: Compound isolation, structure elucidation and bioassays; D.C.W.:

GNPS and SIRIUS analysis. T.D.T., D.W.E., and D.C.W.: Methodology, investigation, data curation, formal analysis, writing—original draft preparation, writing—review and editing. All authors have read and agreed to the published version of the manuscript.

Funding: This research was funded by GCR Primary Industries Pty Ltd.

Institutional Review Board Statement: Not applicable.

Informed Consent Statement: Not applicable.

Data Availability Statement: The original contributions presented in this study are included in the article/Supplementary Materials. The original GNPS data can be found at the following links: Negative MS/MS data: https://gnps.ucsd.edu/ProteoSAFe/status.jsp?task=ee158423f09a4993b166ac837c5d4204 (accessed on 15 August 2024). Positive MS/MS data: https://gnps.ucsd.edu/ProteoSAFe/status.jsp?task=ee158423f09a4993b166ac837c5d4204 (accessed on 15 August 2024). Further inquiries can be directed to the corresponding author.

Acknowledgments: The authors thank Gye Duncan, who donated a propolis sample to this study. We thank GCR Primary Industries Pty Ltd. for financial support. We acknowledge the Australian Research Council for support towards an NMR instrument (ARC LE140100119) and the UniSC GROW for support towards an uHPLC-QTOF MS system. This work is supported by Galaxy Australia, a service provided by Australian BioCommons and its partners. The service receives NCRIS funding through Bioplatforms Australia and the University of Melbourne and Queensland Government RICF funding.

Conflicts of Interest: The authors declare no conflicts of interest.

References

1. Al-Hatamleh, M.A.I.; Boer, J.C.; Wilson, K.L.; Plebanski, M.; Mohamud, R.; Mustafa, M.Z. Antioxidant-Based Medicinal Properties of Stingless Bee Products: Recent Progress and Future Directions. *Biomolecules* **2020**, *10*, 923. [CrossRef] [PubMed]
2. Ghisalberti, E.L. Propolis: A review. *Bee World* **1979**, *60*, 59–84. [CrossRef]
3. Leonhardt, S.D.; Blüthgen, N. A Sticky Affair: Resin Collection by Bornean Stingless Bees. *Biotropica* **2009**, *41*, 730–736. [CrossRef]
4. Bankova, V.; Popova, M.; Trusheva, B. The phytochemistry of the honeybee. *Phytochemistry* **2018**, *155*, 1–11. [CrossRef]
5. Popova, M.; Trusheva, B.; Bankova, V. Propolis of stingless bees: A phytochemist's guide through the jungle of tropical biodiversity. *Phytomedicine* **2021**, *86*, 153098. [CrossRef] [PubMed]
6. Zulhendri, F.; Perera, C.O.; Chandrasekaran, K.; Ghosh, A.; Tandean, S.; Abdulah, R.; Herman, H.; Lesmana, R. Propolis of stingless bees for the development of novel functional food and nutraceutical ingredients: A systematic scoping review of the experimental evidence. *J. Funct. Foods* **2022**, *88*, 104902. [CrossRef]
7. Tran, T.D.; Ogbourne, S.M.; Brooks, P.R.; Sánchez-Cruz, N.; Medina-Franco, J.L.; Quinn, R.J. Lessons from Exploring Chemical Space and Chemical Diversity of Propolis Components. *Int. J. Mol. Sci.* **2020**, *21*, 4988. [CrossRef]
8. Reynolds, O.L.; Robinson, M. *Australian Native Bee Strategic RD&E Plan (2022–2027)*; AgriFutures Australia Publication: Wagga Wagga, NSW, Australia, 2022; p. 40.
9. Halcroft, M.; Spooner-Hart, R.; Dollin, L.A. Australian Stingless Bees. In *Pot-Honey: A Legacy of Stingless Bees*; Vit, P., Pedro, S.R.M., Roubik, D., Eds.; Springer: New York, NY, USA, 2013; pp. 35–72. [CrossRef]
10. Massaro, F.C.; Brooks, P.R.; Wallace, H.M.; Russell, F.D. Cerumen of Australian stingless bees (*Tetragonula carbonaria*): Gas chromatography-mass spectrometry fingerprints and potential anti-inflammatory properties. *Naturwissenschaften* **2011**, *98*, 329–337. [CrossRef]
11. Massaro, F.C.; Brooks, P.R.; Wallace, H.M.; Nsengiyumva, V.; Narokai, L.; Russell, F.D. Effect of Australian Propolis from Stingless Bees (*Tetragonula carbonaria*) on Pre-Contracted Human and Porcine Isolated Arteries. *PLoS ONE* **2013**, *8*, e81297. [CrossRef]
12. Massaro, C.F.; Katouli, M.; Grkovic, T.; Vu, H.; Quinn, R.J.; Heard, T.A.; Carvalho, C.; Manley-Harris, M.; Wallace, H.M.; Brooks, P. Anti-staphylococcal activity of C-methyl flavanones from propolis of Australian stingless bees (*Tetragonula carbonaria*) and fruit resins of *Corymbia torelliana* (Myrtaceae). *Fitoterapia* **2014**, *95*, 247–257. [CrossRef]
13. Nishimura, E.; Murakami, S.; Suzuki, K.; Amano, K.; Tanaka, R.; Shinada, T. Structure Determination of Monomeric Phloroglucinol Derivatives with a Cinnamoyl Group Isolated from Propolis of the Stingless Bee, *Tetragonula carbonaria. Asian J. Org. Chem.* **2016**, *5*, 855–859. [CrossRef]

14. Hamilton, K.D.; Czajkowski, D.; Kong, N.J.; Tran, T.D.; Gustafson, K.R.; Pauly, G.; Boyle, G.M.; Simmons, J.L.; Steadman, R.; Moseley, R.; et al. Anti-Fibrotic Potential of Tomentosenol A, a Constituent of Cerumen from the Australian Native Stingless Bee, *Tetragonula carbonaria*. *Antioxidants* **2022**, *11*, 1604. [CrossRef] [PubMed]

15. Reddy, V.P. Oxidative Stress in Health and Disease. *Biomedicines* **2023**, *11*, 2925. [CrossRef]

16. Bondonno, N.P.; Dalgaard, F.; Murray, K.; Davey, R.J.; Bondonno, C.P.; Cassidy, A.; Lewis, J.R.; Kyrø, C.; Gislason, G.; Scalbert, A.; et al. Higher Habitual Flavonoid Intakes Are Associated with a Lower Incidence of Diabetes. *J. Nutr.* **2021**, *151*, 3533–3542. [CrossRef]

17. Tomic, D.; Chen, L.; Moran, L.L.; Magliano, D.J.; Shaw, J.E. Causes of death among Australians with type 1 or type 2 diabetes, 2002–2019. *Diabet. Med.* **2024**, *41*, e15206. [CrossRef] [PubMed]

18. Dong, Y.; Sui, L.; Yang, F.; Ren, X.; Xing, Y.; Xiu, Z. Reducing the intestinal side effects of acarbose by baicalein through the regulation of gut microbiota: An in vitro study. *Food Chem.* **2022**, *394*, 133561. [CrossRef]

19. Vinayagam, R.; Jayachandran, M.; Xu, B. Antidiabetic Effects of Simple Phenolic Acids: A Comprehensive Review. *Phytother. Res.* **2016**, *30*, 184–199. [CrossRef]

20. Pieczykolan, A.; Pietrzak, W.; Gawlik-Dziki, U.; Nowak, R. Antioxidant, Anti-Inflammatory, and Anti-Diabetic Activity of Phenolic Acids Fractions Obtained from *Aerva lanata* (L.) Juss. *Molecules* **2021**, *26*, 3486. [CrossRef]

21. Tran, C.T.N.; Brooks, P.R.; Bryen, T.J.; Williams, S.; Berry, J.; Tavian, F.; McKee, B.; Tran, T.D. Quality assessment and chemical diversity of Australian propolis from Apis mellifera bees. *Sci. Rep.* **2022**, *12*, 13574. [CrossRef]

22. Wang, M.; Carver, J.J.; Phelan, V.V.; Sanchez, L.M.; Garg, N.; Peng, Y.; Nguyen, D.D.; Watrous, J.; Kapono, C.A.; Luzzatto-Knaan, T.; et al. Sharing and community curation of mass spectrometry data with Global Natural Products Social Molecular Networking. *Nat. Biotechnol.* **2016**, *34*, 828–837. [CrossRef]

23. Uddin, S.; Brooks, P.R.; Tran, T.D. Chemical Characterization, α-Glucosidase, α-Amylase and Lipase Inhibitory Properties of the Australian Honey Bee Propolis. *Foods* **2022**, *11*, 1964. [CrossRef] [PubMed]

24. The Galaxy, C. The Galaxy platform for accessible, reproducible, and collaborative data analyses: 2024 update. *Nucleic Acids Res.* **2024**, *52*, W83–W94. [CrossRef] [PubMed]

25. Tukiran, B.; Mahmudah, F.; Hidayati, N.; Shimizu, K. A phenolic acid and its antioxidant activity from stem bark of chloroform fraction of *Syzygium littorale* (Blume) amshoff (Myrtaceae). *Molekul* **2016**, *11*, 180–189. [CrossRef]

26. Davis, A.L.; Cai, Y.; Davies, A.P.; Lewis, J.R. 1H and 13C NMR Assignments of Some Green Tea Polyphenols. *Magn. Reson. Chem.* **1996**, *34*, 887–890. [CrossRef]

27. Park, S.Y.; Bae, Y.S. Antioxidative Activity of *Prunus sargentii* Outer Bark Extractives. *J. Korean Wood Sci. Technol.* **2012**, *40*, 141–146. [CrossRef]

28. He, D.; Gu, D.; Huang, Y.; Ayupbek, A.; Yang, Y.; Aisa, H.A.; Ito, Y. Separation and Purification of Phenolic Acids and Myricetin from Black Currant by High-Speed Countercurrent Chromatography. *J. Liq. Chromatogr. Relat. Technol.* **2009**, *32*, 3077–3088. [CrossRef]

29. Chu, L.L.; Pandey, R.P.; Jung, N.; Jung, H.J.; Kim, E.-H.; Sohng, J.K. Hydroxylation of diverse flavonoids by CYP450 BM3 variants: Biosynthesis of eriodictyol from naringenin in whole cells and its biological activities. *Microb. Cell Factories* **2016**, *15*, 135. [CrossRef]

30. Granados-Pineda, J.; Uribe-Uribe, N.; García-López, P.; Ramos-Godinez, M.D.; Rivero-Cruz, J.F.; Pérez-Rojas, J.M. Effect of Pinocembrin Isolated from Mexican Brown Propolis on Diabetic Nephropathy. *Molecules* **2018**, *23*, 852. [CrossRef]

31. Malterud, K.E. C-methylated dihydrochalcones from Myrica gale fruit exudate. *Acta Pharm. Nord.* **1992**, *4*, 65–68.

32. Wollenweber, E.; Wehde, R.; Dörr, M.; Lang, G.; Stevens, J.F. C-Methyl-flavonoids from the leaf waxes of some Myrtaceae. *Phytochemistry* **2000**, *55*, 965–970. [CrossRef]

33. Freitas, M.O.; Ponte, F.A.F.; Lima, M.A.S.; Silveira, E.R. Flavonoids and triterpenes from the nest of the stingless bee Trigona spinipes. *J. Braz. Chem. Soc.* **2008**, *19*, 532–535. [CrossRef]

34. Heim, K.E.; Tagliaferro, A.R.; Bobilya, D.J. Flavonoid antioxidants: Chemistry, metabolism and structure-activity relationships. *J. Nutr. Biochem.* **2002**, *13*, 572–584. [CrossRef] [PubMed]

35. Department of Agriculture, Water and the Environment. Conservation Advice for *Melichrus* sp. Gibberagee (Narrow-Leaf Melichrus). Department of Agriculture, Water and the Environment, The Australian Government Canberra, 2021. Available online: https://www.environment.gov.au/biodiversity/threatened/species/pubs/86881-conservation-advice-23112021.pdf (accessed on 24 February 2025).

36. Samson, S.L.; Garber, A.J. Prevention of type 2 Diabetes Mellitus: Potential of pharmacological agents. *Best Pract. Res. Clin. Endocrinol. Metab.* **2016**, *30*, 357–371. [CrossRef]

37. Gao, H.; Nishioka, T.; Kawabata, J.; Kasai, T. Structure-activity relationships for alpha-glucosidase inhibition of baicalein, 5,6,7-trihydroxyflavone: The effect of A-ring substitution. *Biosci. Biotechnol. Biochem.* **2004**, *68*, 369–375. [CrossRef] [PubMed]

38. Proença, C.; Freitas, M.; Ribeiro, D.; Oliveira, E.F.T.; Sousa, J.L.C.; Tomé, S.M.; Ramos, M.J.; Silva, A.M.S.; Fernandes, P.A.; Fernandes, E. α-Glucosidase inhibition by flavonoids: An in vitro and in silico structure–activity relationship study. *J. Enzym. Inhib. Med. Chem.* **2017**, *32*, 1216–1228. [CrossRef] [PubMed]

39. Kumar, S.; Pandey, A.K. Chemistry and biological activities of flavonoids: An overview. *Sci. World J.* **2013**, *2013*, 162750. [CrossRef]

40. Yang, H.; Dong, Y.; Du, H.; Shi, H.; Peng, Y.; Li, X. Antioxidant Compounds from Propolis Collected in Anhui, China. *Molecules* **2011**, *16*, 3444–3455. [CrossRef]

41. Jasemi, S.V.; Khazaei, H.; Morovati, M.R.; Joshi, T.; Aneva, I.Y.; Farzaei, M.H.; Echeverría, J. Phytochemicals as treatment for allergic asthma: Therapeutic effects and mechanisms of action. *Phytomedicine* **2024**, *122*, 155149. [CrossRef]

42. Jasemi, S.V.; Khazaei, H.; Momtaz, S.; Farzaei, M.H.; Echeverría, J. Natural products in the treatment of pulmonary emphysema: Therapeutic effects and mechanisms of action. *Phytomedicine* **2022**, *99*, 153988. [CrossRef]

43. Kahkeshani, N.; Farzaei, F.; Fotouhi, M.; Alavi, S.S.; Bahramsoltani, R.; Naseri, R.; Momtaz, S.; Abbasabadi, Z.; Rahimi, R.; Farzaei, M.H.; et al. Pharmacological effects of gallic acid in health and diseases: A mechanistic review. *Iran. J. Basic Med. Sci.* **2019**, *22*, 225–237.

44. Hadidi, M.; Liñán-Atero, R.; Tarahi, M.; Christodoulou, M.C.; Aghababaei, F. The Potential Health Benefits of Gallic Acid: Therapeutic and Food Applications. *Antioxidants* **2024**, *13*, 1001. [CrossRef]

45. Obafemi, T.O.; Jaiyesimi, K.F.; Olomola, A.A.; Olasehinde, O.R.; Olaoye, O.A.; Adewumi, F.D.; Afolabi, B.A.; Adewale, O.B.; Akintayo, C.O.; Ojo, O.A. Combined effect of metformin and gallic acid on inflammation, antioxidant status, endoplasmic reticulum (ER) stress and glucose metabolism in fructose-fed streptozotocin-induced diabetic rats. *Toxicol. Rep.* **2021**, *8*, 1419–1427. [CrossRef] [PubMed]

46. Bai, J.; Zhang, Y.; Tang, C.; Hou, Y.; Ai, X.; Chen, X.; Zhang, Y.; Wang, X.; Meng, X. Gallic acid: Pharmacological activities and molecular mechanisms involved in inflammation-related diseases. *Biomed. Pharmacother.* **2021**, *133*, 110985. [CrossRef]

47. Bernatoniene, J.; Kopustinskiene, D.M. The Role of Catechins in Cellular Responses to Oxidative Stress. *Molecules* **2018**, *23*, 965. [CrossRef]

48. Ito, H.; Shoji, Y.; Matsumoto, K.-i.; Fukuhara, K.; Nakanishi, I. Anti-cancer Effect of a Planar Catechin Analog through the Decrease in Mitochondrial Membrane Potential. *ACS Med. Chem. Lett.* **2023**, *14*, 1478–1481. [CrossRef] [PubMed]

49. Stefaniu, A.; Pirvu, L.C. In Silico Study Approach on a Series of 50 Polyphenolic Compounds in Plants; A Comparison on the Bioavailability and Bioactivity Data. *Molecules* **2022**, *27*, 1413. [CrossRef]

50. Fan, F.Y.; Sang, L.X.; Jiang, M. Catechins and Their Therapeutic Benefits to Inflammatory Bowel Disease. *Molecules* **2017**, *22*, 484. [CrossRef]

51. Gopal, J.; Muthu, M.; Paul, D.; Kim, D.-H.; Chun, S. Bactericidal activity of green tea extracts: The importance of catechin containing nano particles. *Sci. Rep.* **2016**, *6*, 19710. [CrossRef] [PubMed]

52. Ayuda-Durán, B.; Garzón-García, L.; González-Manzano, S.; Santos-Buelga, C.; González-Paramás, A.M. Insights into the Neuroprotective Potential of Epicatechin: Effects against Aβ-Induced Toxicity in *Caenorhabditis elegans*. *Antioxidants* **2024**, *13*, 79. [CrossRef]

53. Taheri, Y.; Suleria, H.A.R.; Martins, N.; Sytar, O.; Beyatli, A.; Yeskaliyeva, B.; Seitimova, G.; Salehi, B.; Semwal, P.; Painuli, S.; et al. Myricetin bioactive effects: Moving from preclinical evidence to potential clinical applications. *BMC Complement. Med. Ther.* **2020**, *20*, 241. [CrossRef]

54. Islam, A.; Islam, M.S.; Rahman, M.K.; Uddin, M.N.; Akanda, M.R. The pharmacological and biological roles of eriodictyol. *Arch. Pharmacal Res.* **2020**, *43*, 582–592. [CrossRef] [PubMed]

55. Elbatreek, M.H.; Mahdi, I.; Ouchari, W.; Mahmoud, M.F.; Sobeh, M. Current advances on the therapeutic potential of pinocembrin: An updated review. *Biomed. Pharmacother.* **2023**, *157*, 114032. [CrossRef] [PubMed]

56. Vechi, G.; da Silva, R.d.C.M.V.d.A.F.; de Souza, P.; da Silva, L.M.; de Andrade, S.F.; Cechinel Filho, V. Cryptostrobin and catechin isolated from Eugenia mattosii D. Legrand leaves induce endothelium-dependent and independent relaxation in spontaneously hypertensive rat aorta. *Pharmacol. Rep.* **2019**, *71*, 950–957. [CrossRef] [PubMed]

57. Oikawa, N.; Nobushi, Y.; Wada, T.; Sonoda, K.; Okazaki, Y.; Tsutsumi, S.; Park, Y.K.; Kurokawa, M.; Shimba, S.; Yasukawa, K. Inhibitory effects of compounds isolated from the dried branches and leaves of murta (*Myrceugenia euosma*) on lipid accumulation in 3T3-L1 cells. *J. Nat. Med.* **2016**, *70*, 502–509. [CrossRef]

58. Gafner, S.; Wolfender, J.L.; Mavi, S.; Hostettmann, K. Antifungal and antibacterial chalcones from *Myrica serrata*. *Planta Medica* **1996**, *62*, 67. [CrossRef]

59. Abdel Bar, F.M. Genus Melaleuca—A Review on the Phytochemistry and Pharmacological Activities of the Non-Volatile Components. *Rec. Nat. Prod.* **2021**, *15*, 219–242. [CrossRef]

60. Cuong, N.M.; Khanh, P.N.; Nhung, L.T.H.; Ha, N.X.; Huong, T.T.; Bauerova, K.; Kim, Y.H.; Tung, D.D.; Thuy, T.T.; Anh, N.T.H. Acetylcholinesterase inhibitory activities of some flavonoids from the root bark of Pinus krempfii Lecomte: In vitro and in silico study. *J. Biomol. Struct. Dyn.* **2024**, *42*, 4888–4901. [CrossRef]

61. Tarawneh, A.H.; Ibrahim, M.A.; Radwan, M.M.; Ma, G.; Cutler, H.G.; Cutler, S.J. Therapeutic Efficacy of Phenolic Compounds from Micnia prasina. in Acute Lymphoblastic Leukemia. *Planta Medica* **2013**, *79*, P55. [CrossRef]
62. Stompor, M. A Review on Sources and Pharmacological Aspects of Sakuranetin. *Nutrients* **2020**, *12*, 513. [CrossRef]
63. Uçar, K.; Göktaş, Z. Biological activities of naringenin: A narrative review based on in vitro and in vivo studies. *Nutr. Res.* **2023**, *119*, 43–55. [CrossRef]
64. Norkaew, C.; Subkorn, P.; Chatupheeraphat, C.; Roytrakul, S.; Tanyong, D. Pinostrobin, a fingerroot compound, regulates miR-181b-5p and induces acute leukemic cell apoptosis. *Sci. Rep.* **2023**, *13*, 8084. [CrossRef] [PubMed]
65. Zhao, L.-L.; Jayeoye, T.J.; Ashaolu, T.J.; Olatunji, O.J. Pinostrobin, a dietary bioflavonoid exerts antioxidant, anti-inflammatory, and anti-apoptotic protective effects against methotrexate-induced ovarian toxicity in rats. *Tissue Cell* **2023**, *85*, 102254. [CrossRef] [PubMed]
66. Kazmi, I.; Al-Abbasi, F.A.; Afzal, M.; Shahid Nadeem, M.; Altayb, H.N. Sterubin protects against chemically-induced Alzheimer's disease by reducing biomarkers of inflammation- IL-6/ IL-β/ TNF-α and oxidative stress- SOD/MDA in rats. *Saudi J. Biol. Sci.* **2023**, *30*, 103560. [CrossRef] [PubMed]

Article

Chemical and Functional Characterization of Propolis Collected from Different Areas of South Italy

Giulia Grassi [1,*], Giambattista Capasso [2], Emilio Gambacorta [2] and Anna Maria Perna [2]

[1] Department of Agricultural, Environmental and Food Sciences, University of Molise, Via De Sanctis 1, 86100 Campobasso, Italy

[2] School of Agricultural, Forestry, Food and Environmental Sciences, University of Basilicata, Viale dell'Ateneo Lucano 10, 85100 Potenza, Italy; giambattista.capasso@unibas.it (G.C.); emilio.gambacorta@unibas.it (E.G.); anna.perna@unibas.it (A.M.P.)

* Correspondence: giulia.grassi@unimol.it; Tel.: +39-0971-205073; Fax: +39-0971-205099

Abstract: This study investigated the chemical and functional characterization of propolis collected in southern Italy, in particular in Basilicata, a region rich in ecological and vegetative biodiversity. Sixteen samples of propolis, collected within a radius of 40 km from each other in the Basilicata region, showed significant differences between the chemical and functional parameters investigated: color index (L*, a*, b*; $p < 0.05$) and variation in chemical composition and antioxidant activities by ABTS and FRAP assays. In general, Lucanian propolis had a low content of waxes ($p < 0.05$) and a high content of resin ($p < 0.05$) and balsams ($p < 0.05$). The content of the total phenolic compounds and flavonoids was highly variable, as was the biological capacity. In conclusion, Lucanian propolis showed remarkable variability, highlighting significant diversification according to the geographical position and the diversity of the flora surrounding the apiary that the bees use as a source of resin. This study, therefore, contributes to the enhancement of the quality of propolis, laying the foundations for the production and marketing of propolis not only in the food industry but also in the pharmaceutical and cosmetic industries.

Keywords: propolis; chemical and functional composition; antioxidant activity; Basilicata

Citation: Grassi, G.; Capasso, G.; Gambacorta, E.; Perna, A.M. Chemical and Functional Characterization of Propolis Collected from Different Areas of South Italy. *Foods* **2023**, *12*, 3481. https://doi.org/10.3390/foods12183481

Academic Editors: Liming Wu and Qiangqiang Li

Received: 29 August 2023
Revised: 13 September 2023
Accepted: 14 September 2023
Published: 19 September 2023

1. Introduction

Propolis is a natural product that is processed by bees with the addition of other substances such as wax, pollen and glandular secretions. It appears as a resinous substance that arises from the industrious work of foraging bees that feed on substances present on the bark and exudates of numerous plants [1,2].

Propolis is a product that is used inside the hive to protect and build borders and entrances for bees but also as an insulator thanks to its ability to be sticky in hot periods and rigid in cold periods. It is commonly called "bees glue" and it is an excellent biological defense against the proliferation of microorganisms [3,4]. Furthermore, bees use propolis as a natural remedy to prevent the decomposition of the carcass of other insects that settle in the hives and it is able to stabilize the internal temperature of the hive at around 35/37 °C. Finally, being a lipophilic substance, it prevents the penetration of water into the hive with a consequent stabilization of humidity and regulates the airflow in the hive [2]. In nature, there are numerous types of propolis that differ from each other in chemical composition, physical characteristics, color and other characteristics that make them unique. In fact, some researchers have noticed differences between various propolis samples in consistency; some were brittle and hard, while others were elastic and gummy. In general, propolis is characterized by 50% resin (phenolic compounds) and vegetable balms, 30% wax, 10% essential and aromatic oils, 5% pollen and 5% other substances, including also organic debris [5]. In reality, there are a number of factors that influence the composition of raw propolis: geographical area, botanical origin, seasonality, climatic temperatures and others. Also, the color of propolis is influenced by the geographical area and the plant source on which the bees feed [5]. The components that characterize this product are numerous: steroids, amino acids, phenols, terpenes, flavonoids, carbohydrates, aliphatic and aromatic acids and esters. Its use in cosmetic, pharmaceutical and food preparations is also increasing, thanks to the therapeutic, preventive and improvement activities of our body [6]. The biological effects of this matrix are a broad spectrum: antibacterial and antioxidant, antitumor, cardioprotective, antiviral, immunomodulatory, hepatoprotective, neuroprotective, antidiabetic, anti-inflammatory, anesthetic and antiallergic due to the abundant mineral content [7–9]. The need to chemically type propolis could be useful in order to officially include it within the health system, so as to guarantee the safety and quality of propolis in health and therapeutic fields. The chemical composition of propolis depends on the geographical and climatic characteristics of the place of collection, which makes propolis an extremely variable matrix and, consequently, makes it difficult to standardize the characteristics of propolis. Therefore, it is necessary to investigate its chemical composition and also the relative biological properties in order to contribute to enriching the partially already existing knowledge. In support, Graikou et al. [10] have demonstrated the presence of different propolis in the Mediterranean area in relation to its geographical origin. Gardini et al. [11] highlighted the need to investigate the variability of the composition of propolis collected in the Italian territory, according to the ecoregion of origin suggested by Blasi et al. [12]. Our study is inserted within this context and aims to quantify the physicochemical parameters, color index, total phenolic content and flavonoids and evaluate the antioxidant activity. In addition, the correlations between the parameters analyzed in propolis of different geographical origins of southern Italy were also studied, with particular reference to the Basilicata region, a region rich in ecological and vegetative biodiversity.

2. Materials and Methods

2.1. Chemicals and Apparatus

The 2,2′-azino-bis-(3-ethylbenzothiazoline-6-sulfonic acid) (ABTS), 2,4,6-tripyridyl-s-triazine (TPTZ), potassium persulfate, hydrochloric acid, ferric chloride, iron(II) sulfate heptahydrate, sodium phosphate, sodium hydroxide and ammonium persulfate were purchased from Sigma-Aldrich (Milan, Italy). The phenolic compounds, gallic acid and quercetin, were purchased from Sigma Chemical Co. (St Louis, MO, USA). Analytical grade reagents, such as sodium carbonate, potassium hydroxide, Folin-Ciocalteu reagent, ethanol, methanol and hexane were obtained from Panreac (Barcelona, Spain). Aluminum chloride, potassium ferricyanide, ferric chloride and trichloroacetic acid were from Sigma Chemical Co. (St. Louis, MO, USA). The water was treated in a Milli-Q water purification system (TGI Pure Water Systems, Brea, CA, USA). Coomassie Brilliant blue G250 was purchased from Bio-Rad (Richmond, CA, USA). The spectrophotometer UV-VIS Spectrophotometer 1204 (Shimadzu, Japan) was used. MINOLTA Chromameter CR-300 (Minolta Camera Corp., Meter Division, Ramsey, NJ, USA) equipped with a D65 illuminant, 10° Observer and zero and white calibration was used to measure the color parameters (CIE L*, a*, b*).

2.2. Propolis Samples

A total number of 16 propolis samples were taken directly from hives located in 8 different areas of the Basilicata region (Italy), according to availability and beekeeping activity, collected by individual beekeepers during the 2022 harvest (Figure 1). These areas are differentiated by differences in geographical position, climatic-environmental factors and soil composition, showing a different ecological-vegetative climate and great biodiversity. Propolis samples were randomly obtained after honey extraction by conventional scraping. After removal of debris, propolis samples were stored at −20 °C until analysis.

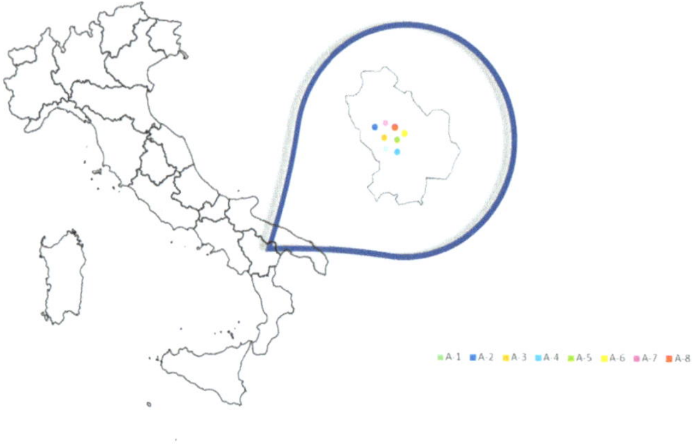

Figure 1. Distribution of propolis sampling areas in the 8 areas in the Basilicata region. Each area is 30 to 40 km apart from each other.

2.3. Water Content

In order to determine the free water content, it was determined following the protocol suggested by Funari et al. [13]. The sample was dried in a conventional oven at 105 °C for 2 h until constant weight was reached.

2.4. Ash Content

The ash content was determined following the AOAC [14] procedure, by ashing the raw propolis samples at 600 °C.

2.5. Wax Extraction

The wax contents were estimated according to a procedure described by Papotti et al. [15]. Three grams of frozen propolis was powdered and treated with 120 mL of petroleum ether at 40−60 °C in a Soxhlet extractor for 6 h. The extract was transferred to a previously weighed 150 mL evaporator flask and concentrated under reduced pressure at 50 °C. Then, 120 mL of 70% ethanol was added, heated under reflux until a clear solution was obtained and then cooled at 0 °C for 1 h to promote wax separation. The mixture was filtered through a previously weighed Whatman grade no. 41 filter paper. The flask and the filter were washed with 70% ethanol, dried at 110 °C for 1 h and transferred to a desiccator until constant weight. The sum of the residues remaining in the flask and on the filter, expressed as % w/w, represents the waxes.

2.6. Balsam Extraction and Quantification

The contents of balsams were estimated according to a procedure described by Papotti et al. [15]. The 70% ethanolic filtrate obtained during wax extraction was concentrated under reduced pressure at 60 °C. The aqueous residue was transferred to a separating funnel and 50 mL of dichlorometane was added. After shaking, the organic phase was collected and dried over 30 g of anhydrous Na_2SO_4 and then filtered in a previously weighed 150 mL evaporator flask. The extraction was repeated twice. The solution was evaporated to dryness under reduced pressure at 60 °C and the flask was transferred to a desiccator until constant weight. The results are expressed as % w/w.

2.7. Resin Extraction

The contents of resins were estimated according to a procedure described by Papotti et al. [15]. The residual propolis obtained after the extraction in the Soxhlet equipment was treated with 120 mL of a mixture of chloroform/ethanol 1:1 (v/v) in a Soxhlet extractor for 6 h. The extract was transferred to a preweighed 150 mL evaporator flask and concentrated to dryness under reduced pressure at 70 °C. The flask was dried at 110 °C for 1 h and transferred to a desiccator until constant weight. The results are expressed as % w/w.

2.8. Colorimetric Analysis

To determine the color indices of the propolis samples, the following were recorded: L* (lightness), a* (redness-green) and b* (yellow-blue). The colorimeter was previously calibrated using a standard white plate (L* = 94.56, a* = -0.31, b* = 4.16, C*ab = 4.18). The analysis was performed in quadruplicate.

2.9. Extraction of Phenolic Compounds

Pretreatment was required to determine the total phenol content and the flavonoid content. The method was suggested by Özkök et al. [16], with slight modifications. The sample was mixed with 75% ethanol/water (v/v), homogenized and sonicated in an ultrasonic bath for 5 h and finally centrifuged.

2.10. Total Phenols Content (TPC)

The total phenolic content was determined by a modification of the Folin–Ciocalteu method, as described by Escheriche et al. [17], with some modifications. A volume of ethanolic extracts (500 μL) were mixed with 250 μL of Folin–Ciocalteu reagent. After 3 min, 1000 μL saturated sodium carbonate solution was added to the mixture. The solution was then incubated at room temperature for 1 h and the absorbance was measured at 760 nm. The gallic acid calibration curve was used to determine the total phenolic content and the results were expressed as mg gallic acid equivalent per g propolis (0.0125 to 0.1 mg/mL). Analyses were performed in triplicate.

2.11. Total Flavonoid Content (TFC)

The total flavonoid content of the crude extract was determined by the aluminum chloride colorimetric method as suggested by Escheriche et al. [17], with some modifications. A volume of supernatant (50 μL) was mixed with 1500 μL of 2% aluminum chloride in methanol and 1350 μL of methanol. After 30 min of incubation in the dark at room temperature, the absorbance at 415 nm was measured. The blank test replaced the sample with distilled water and a volume of 2850 μL of methanol, placed under the same incubation conditions. Quercitin was used to calculate the standard curve (0.02 to 0.25 mg/mL) and the results were expressed as mg quercitin equivalents per g propolis. Analyses were performed in triplicate.

2.12. Antioxidant Power (FRAP) Method

The FRAP assay was conducted following the method described by Chaves et al. [18], with some modifications. The extract (200 μL) was mixed with 2800 μL of FRAP reagent. This reagent was previously prepared by mixing 300 mM sodium acetate buffer solution at pH 3.6, 10 mM TPZT and 20 mM FeCl3 hexahydrate, in a ratio of 10:1:1, respectively. The mixture was incubated for 30 min at 37 °C and subsequently read at 593 nm. The blank was prepared by replacing the same amount of diluted extract with methanol. The results were expressed in mM equivalents of Trolox per gram of propolis, after performing a calibration curve at known concentrations (0.01 to 0.1 mM of Trolox/mL). Analyses were performed in triplicate.

2.13. ABTS Free Radical Scavenging

The antioxidant activity of propolis extracts by the ABTS spectrophotometric assay was determined with the method suggested by Chaves et al. [18], with some modifications. The extract (100 μL) was mixed with 2900 μL of the ABTS$^+$ dilution. The decrease in absorbance at 734 nm was measured after 30 min of incubation at room temperature. The blank was prepared with methanol only. The absorbance was read at 730 nm after 30 min.

The results were expressed in mM equivalents of Trolox per gram of propolis. Analyses were performed in triplicate.

2.14. Statistical Analysis

Statistical analysis was performed using the general linear model (GLM) procedure of statistical analysis system SAS [19], using a monofactorial model: $y_{ik} = \mu + \alpha_i + \varepsilon_{ik}$; where: μ = average mean; α_i = effect of geographical origin (1, ..., 8); and ε_{ik} = experimental error. The Student's *t*-test was used for all variables comparisons. Differences between means at the 95% ($p < 0.05$) confidence level were considered statistically significant. Pearson correlation coefficient (r) was used to analyze the correlations between different parameters of propolis samples.

3. Results

3.1. Physico-Chemical Composition in Propolis Samples

The physicochemical parameters of propolis samples collected in eight geographic areas of the Basilicata region are summarized in Table 1.

Table 1. Physico-chemical composition (%) in fresh propolis differently by collection area.

	Dry Matter		Ash		Wax		Resin		Balsam	
	μ	^{1} SD	μ	^{1} SD	μ	^{1} SD	μ	^{1} SD	μ	^{1} SD
A-1	3.72 [ab]	0.31	1.75 [a]	0.09	31.55 [a]	1.82	48.67 [a]	1.67	4.33 [a]	0.2
A-2	3.23 [a]	0.30	1.01 [b]	0.07	30.4 [ab]	1.47	56.49 [b]	2.25	6.39 [b]	0.23
A-3	3.39 [ab]	0.18	0.89 [c]	0.05	29.08 [bc]	1.01	59.81 [c]	1.23	6.4 [b]	0.26
A-4	3.79 [b]	0.28	1.84 [d]	0.03	27.88 [c]	1.97	64.3 [d]	1.36	7.82 [c]	0.51
A-5	3.78 [ab]	0.32	0.77 [e]	0.03	27.55 [c]	1.61	63.4 [d]	1.46	4.42 [a]	0.42
A-6	3.31 [a]	0.28	2.06 [f]	0.15	24.47 [d]	1.57	71.78 [e]	1.51	9.63 [d]	0.73
A-7	3.53 [ab]	0.25	0.7 [g]	0.03	30.9 [a]	1.45	58.38 [bc]	1.75	5.41 [e]	0.52
A-8	3.82 [ab]	0.27	0.89 [c]	0.05	24.4 [d]	1.15	68.39 [f]	1.38	4.54 [a]	0.44

^{1} SD standard deviation; [a,b,c,d,e,f,g] means within a column with different superscripts differ ($p < 0.01$).

A statistically significant effect of geographic origin on the physicochemical parameters of propolis from eight different areas in one-way ANOVA was confirmed for all the parameters analyzed ($p < 0.05$), in agreement with Kasote et al. [5]. The moisture content of the propolis provides information on the quality of the propolis; the high water content in propolis indicates improper storage and handling conditions. The average percentage moisture content in the samples investigated was 3.57 ± 0.11, in line with what was found in Moroccan propolis by El Menyiy et al. [20] and no significant differences were found between the samples studied. The factors that influence the moisture content of propolis concern both the handling conditions and the duration of storage, which is to be considered a quality parameter, given the presence of a high content of phenols, which deteriorate easily over time. The ash content also highlighted the presence of inorganic minerals, as well as the presence of impurities present in the sample, probably linked to the natural production process that brings with it different materials, such as wood, remains of bees and small pieces of earth with a consequent increase in the level of ashes. Furthermore, determining the ash content is essential to rule out the possibility of adulteration of the propolis samples [21]. In general, the samples analyzed have an average ash content of $1.22 \pm 0.04\%$, and the values vary from 0.68 ± 0.01 to $2.13 \pm 0.21\%$, in agreement with what is reported in one study conducted on propolis samples from Morocco, in which values were recorded between $0.72 \pm 0.02\%$ and $5.01 \pm 0.01\%$ [22]. Furthermore, in other studies on Mexican propolis [23], the ash content ranged from 0.66% to 5.50%, with respect to the differences in the survey areas. Notably, the ash content of A-5 was significantly higher than the other propolis samples ($p \leq 0.05$). The percentage content of the wax, resin and balsam in the propolis samples under study are shown in Table 1. An average wax content of 27.98 ± 1.68% was recorded in line with what was found by Gardini et al. [11] in Italian propolis. The highest average content among our samples was found to be 31.6 ± 2.84% for sample A-1, while propolis A-6 had the lowest value (23.66 ± 1.71%, $p < 0.05$). The differences found in the wax content could be related to the collection method rather than being influenced by the botanical and/or geographical origin of the sample. The presence of a high content of waxes, and biologically inactive components, could lead to a low percentage of pharmacologically active compounds with consequent repercussions on the commercial value of the product. The number of resins and balms in propolis is directly related to the amount of resin collected by bees during grazing. In general, an average resin content of 61.32% was recorded, while, for the conditioner, it was 6.13%. Some variability was observed for the resin and balsam content between the samples, ranging from 48 to 71% and 4.36 to 9.61%, respectively. In the comparison between the samples, A-6 and A-8 had a resin content significantly higher than 65% ($p < 0.05$). Regarding the balsam content, sample A-5 showed the highest content (9.63%; $p < 0.05$), while the lowest content was found in A-1 and A-8 propolis ($p < 0.05$). The resins and balms contain bioactive plant metabolites that perform numerous biological activities, contributing to the defense of the hive. Papotti et al. [15] highlighted the relationship between the chemical composition and health of bee colonies, in particular the level of resins and balsams contained in propolis. Drescher et al. [3] found, in fact, that the resin content was significantly lower in the colonies more resistant to Varroa and, therefore, the bees resistant to the parasitic mite reserved few resources for resin collection, compared to the bees coming from particularly sensitive colonies. Color is a determining physical-chemical parameter in the choice of the product. The colorimetric characteristics of propolis samples from different geographical areas are shown in Figure 2.

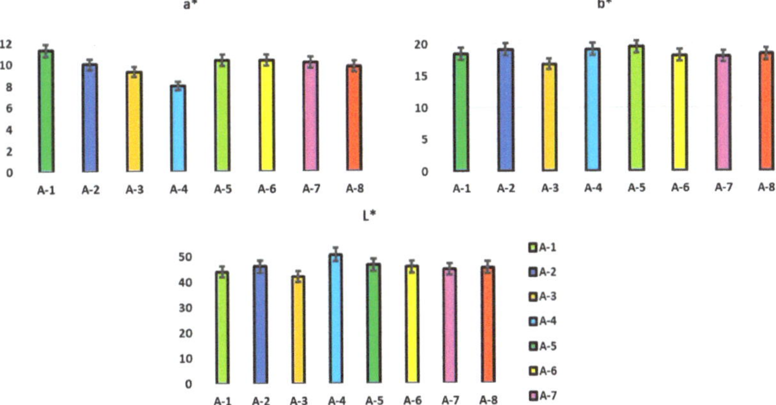

Figure 2. Colorimetric parameters of propolis samples from different geographical areas ($p < 0.05$).

The values, measured with the CIE L*a*b* method, showed high and consistent variability between the propolis samples from different production areas, resulting in being statistically significant ($p < 0.05$). In general, the samples were dark, with L* values ranging from 42.3 to 50.15; in particular, A-3 and A-1 recorded the lowest value of L* (respectively, 42.53 and 43.96), while the brightest were propolis A-4 (L* = 50.15). The parameters a* (red-green) and b* (yellow-blue) of the propolis can be interpreted as a reliable index of the richness in pigments of botanical origin. These values ranged from 7.43 (A-4) to 11.1 (A-1) for parameter a* and from 16.92 (A-3) to 19.45 (A-5) for parameter b*. These values were in line with Portuguese propolis investigated by Gomes et al. [24].

3.2. Content in Total Phenols and Flavonoids

In this study, the content of the total phenols and flavonoids was determined as represented in Figure 3. da Silva et al. [25] suggested that the quality of propolis is based on the content of flavonoid and phenolic compounds since they represent the major bioactive components of propolis, found mainly in resins and balms. The total phenolic compounds ranged between 221 and 461 mg GAE/g propolis (Figure 3).

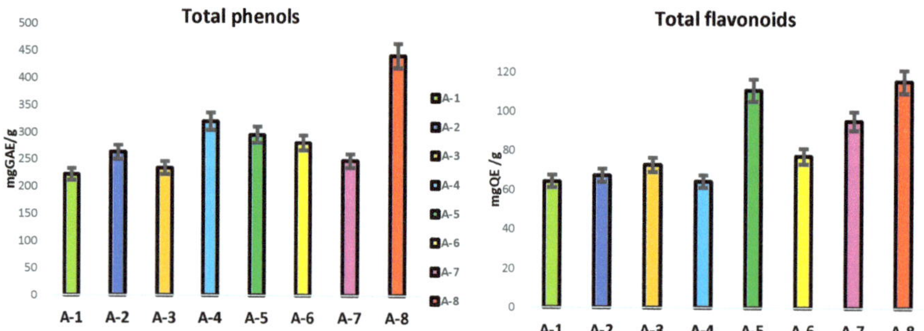

Figure 3. Total phenols (mgGAE/g) and total flavonoids (mgQE/g) of propolis samples from different geographical areas ($p < 0.05$).

Although the samples were collected from areas not far from each other, there were recorded significant differences in their total phenolic content ($p < 0.05$). Propolis A-8 presented the significantly higher total phenol content (442.26 mg GAE/g; $p < 0.01$) while A-1 and A-3 showed the lowest values (222.44 and 234.82 mg GAE/g, respectively). The TPC in propolis extracts from various parts of the world has been extensively studied and a wide range of values can be found in the literature [16]. Turkish propolis reported a range of values of 115–210 mgGAE/g; values from 151 to 329 mgGAE/g were found in Portuguese propolis by Gomes et al. [24]; in Chile and Spain, the recorded values ranged from 200 to 300 mgGAE/g [26], while, in Greek and Cypriot propolis, the range included values from 80 to 338 mgGAE/g [27]. In the analyzed samples, the flavonoid content ranged from 64.35 mgQE/g to 115.62 mgQE/g, and the highest TFC was found for propolis A-8 and A-5 (115.62 and 111.02 mgQE/g, respectively). The lowest TFC was determined for propolis A-1 and A-4 (64.35 and 64.42 mgQE/g, respectively). From our results, it has been demonstrated that the variation is rather limited compared to the data reported in the literature. By way of example, Chinese (8.3–162 mgQE/g) and Australian (0.2–144.8 mgQE/g) propolis observed much wider ranges in total flavonoid content. Furthermore, Kumazawa et al. [28] studied the content of the total flavonoids in propolis from various regions of the world, recording a relatively large variability (2.5–176 mgQE/g). In the literature, very high variability between the studied samples of propolis has been confirmed and the observed differences may derive from various factors: soil composition, temperature, humidity and altitude, which influence the physiological state of the plant and, therefore, on the phenolic biosynthesis.

3.3. Antioxidant Activity in Propolis Samples: ABTS and FRAP

In Figure 4, the antioxidant activities of ABTS and FRAP have been reported. It is well known that propolis exhibits strong antioxidant activity [16] and, in this present work, the ABTS and FRAP assays were chosen for the antioxidant evaluation of the propolis. The ABTS assay highlights the activity of hydrophilic and lipophilic antioxidants, while the FRAP assay uses antioxidants as reducing agents in a redox-linked colorimetric method, employing an easily reduced oxidant system present in stoichiometric excess [29]. Based on antioxidant tests, propolis from different geographic areas has been observed to exhibit varying degrees of antioxidant capacity. The mean values were 5.41 and 1.48 mMTE/g for the ABTS and FRAP assays, respectively. Significant differences have been recorded between the propolis from different locations, suggesting that they have different antioxidant potentials.

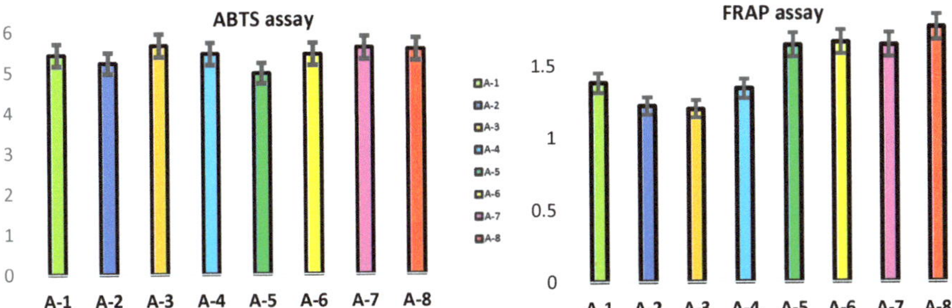

Figure 4. Antioxidant assays (ABTS and FRAP, mMTE/g) of propolis from different geographic areas.

As shown in Figure 4, the values ranged from 4.97 (A-5) to 5.66 mMTE/g (A-3) in the ABTS assay and from 1.2 (A-3) to 1.76 mMTE/g (A-8) in the FRAP assay.

The results obtained by means of the ABTS assay showed a lower antioxidant activity in propolis A-2 and A-5 compared to the propolis of the other areas considered (5.22 and 4.97 mM TE/g, respectively), while the propolis A-3 showed the highest radical scavenging activity (5.66 mM TE/g; $p < 0.05$). The trend of the antioxidant activity, by the FRAP assay, did not confirm the results obtained by the ABTS assay; in particular, the propolis A-3 samples showed the lowest FRAP values (1.2 mM TE/g; $p < 0.05$) than the others, while the maximum activity was observed for propolis sample A-8 (1.76 mMTE/g; $p < 0.01$). Our results showed higher ABTS values than those reported by Martín et al. [30] in Spanish propolis, where they detected values of 1.823 mmol TE/g. The results obtained in this work confirmed and demonstrated that the variations recorded in the antioxidant activity are influenced by the different collection locations, which differ in geographical and climatic factors and in the different composition of the soil. These factors greatly influence the content of biologically active compounds in propolis, which can act synergistically and increase the antioxidant action. A significant and positive linear correlation was observed between FRAP and TPC (r = 0.488; $p < 0.01$) and FRAP and TFC (r = 0.753; $p < 0.01$) in agreement with Kasote et al. [31], who observed how propolis characterized by a high content of phenolic compounds has shown a strong antioxidant activity. In contrast, low and negative correlation coefficients were observed between TPC (r = −0.033) and TFC (r = −0.199) with the ABTS assay; this means that the antioxidant activity of the propolis sample could be due to other non-phenolic components present. Indeed, propolis is characterized by an abundant presence of phytochemicals, including essential oils, minerals and vitamins A, B, C and E, as suggested by Sahlan et al. [32], in which they defined the important and specific role of these components in biological activities. In agreement with our data, propolis from several countries such as Argentina [30], Greece and Cyprus [27], Japan [33] and Poland [33] showed a high correlation between TPC and TFC and the scavenging activity of free radicals. Instead, a negative or absent correlation between them was observed both in the propolis of Morocco and in the propolis of Brazil [22] but also in that of Greece [27]. Our data revealed that propolis with a high content of resin, phenolics and flavonoids had the highest antioxidant activity and that a high amount of flavonoids and phenols was found in samples with a high resin content and low wax content, in line with what was found in Moroccan propolis [20]. However, the number of polyphenols is strongly influenced by the climatic conditions of each collection area, which explains these large differences between the studied samples. Similar data were observed in Brazilian propolis by da Silva et al. [25], in which the relationship between climatic conditions, metabolite profile and antioxidant activity emerged.

4. Discussion

This study summarizes for the first time the state of knowledge on the characteristics of propolis produced in Basilicata and the factors to be considered to characterize the quality of honey.

Overall, it can be concluded that Lucanian propolis has a low wax content and a high content of resins, balms and antioxidant compounds with a marked antioxidant capacity. The chemical, physical, or biological properties of "Lucana" propolis varied considerably between the different propolis according to the geographical location and the diversity of the flora surrounding the apiary that the bees use as a source of resin. Although further work is needed to investigate and define a complete picture of the propolis of the Basilicata region with regard to their chemical composition and therapeutic values, the results of this study provide the basis for the production and commercialization of propolis not only in the food industry but also in the pharmaceutical and cosmetic fields.

Author Contributions: G.G.: methodology, validation, formal analysis, investigation, writing—original draft, writing—review and editing and visualization. G.C.: investigation and methodology. E.G.: methodology, formal analysis and writing—review and editing. A.M.P.: methodology, writing—review and editing, supervision and funding acquisition. All authors have read and agreed to the published version of the manuscript.

Funding: This research received no external funding.

Data Availability Statement: The data that support the findings of this study are available from the corresponding author upon reasonable request.

Conflicts of Interest: The authors declare no conflict of interest.

References

1. Pahlavani, N.; Malekahmadi, M.; Firouzi, S.; Rostami, D.; Sedaghat, A.; Moghaddam, A.B.; Ferns, G.A.; Navashenaq, J.G.; Reazvani, R.; Safarian, M.; et al. Molecular and Cellular Mechanisms of the Effects of Propolis in Inflammation, Oxidative Stress and Glycemic Control in Chronic Diseases. *Nutr. Metab.* **2020**, *17*, 65. [CrossRef] [PubMed]
2. Pant, K.; Thakur, M.; Chopra, H.K.; Dar, B.N.; Nanda, V. Assessment of Fatty Acids, Amino Acids, Minerals, and Thermal Properties of Bee Propolis from Northern India Using a Multivariate Approach. *J. Food Compos. Anal.* **2022**, *111*, 104624. [CrossRef]
3. Drescher, N.; Klein, A.-M.; Neumann, P.; Yañez, O.; Leonhardt, S. Inside Honeybee Hives: Impact of Natural Propolis on the Ectoparasitic Mite Varroa Destructor and Viruses. *Insects* **2017**, *8*, 15. [CrossRef]
4. Simone-Finstrom, M.; Borba, R.S.; Wilson, M.; Spivak, M. Propolis Counteracts Some Threats to Honey Bee Health. *Insects* **2017**, *8*, 46. [CrossRef] [PubMed]
5. Kasote, D.; Bankova, V.; Viljoen, A.M. Propolis: Chemical Diversity and Challenges in Quality Control. *Phytochem. Rev.* **2022**, *21*, 1887–1911. [CrossRef] [PubMed]
6. da Rosa, C.; Bueno, I.L.; Quaresma, A.C.M.; Longato, G.B. Healing Potential of Propolis in Skin Wounds Evidenced by Clinical Studies. *Pharmaceuticals* **2022**, *15*, 1143. [CrossRef] [PubMed]
7. Alanazi, S.; Alenzi, N.; Fearnley, J.; Harnett, W.; Watson, D.G. Temperate Propolis Has Anti-Inflammatory Effects and Is a Potent Inhibitor of Nitric Oxide Formation in Macrophages. *Metabolites* **2020**, *10*, 413. [CrossRef]
8. Santos, L.A.; Rosalen, P.L.; Dias, N.A.; Grisolia, J.C.; Nascimento Gomes, B.J.; Blosfeld-Lopes, L.; Ikegaki, M.; de Alencar, S.M.; Burger, E. Brazilian Red Propolis Shows Antifungal and Immunomodulatory Activities against Paracoccidioides brasiliensis. *J. Ethnopharmacol.* **2021**, *277*, 114181. [CrossRef]
9. Forma, E.; Bryś, M. Anticancer Activity of Propolis and Its Compounds. *Nutrients* **2021**, *13*, 2594. [CrossRef]
10. Graikou, K.; Popova, M.; Gortzi, O.; Bankova, V.; Chinou, I. Characterization and Biological Evaluation of Selected Mediterranean Propolis Samples. Is It a New Type? *LWT-Food Sci. Technol.* **2016**, *65*, 261–267. [CrossRef]
11. Gardini, S.; Bertelli, D.; Marchetti, L.; Graziosi, R.; Pinetti, D.; Plessi, M.; Marcazzan, G.L. Chemical Composition of Italian Propolis of Different Ecoregional Origin. *J. Apic. Res.* **2018**, *57*, 639–647. [CrossRef]
12. Blasi, C.; Capotorti, G.; Copiz, R.; Guida, D.; Mollo, B.; Smiraglia, D.; Zavattero, L. Classification and Mapping of the Ecoregions of Italy. *Plant Biosyst.-Int. J. Deal. Asp. Plant Biol.* **2014**, *148*, 1255–1345. [CrossRef]
13. Funari, C.S.; Ferro, V.O. Análise de Própolis. *Ciênc. Tecnol. Aliment.* **2006**, *26*, 171–178. [CrossRef]
14. AOAC International. Official Methods of Analysis Program. Available online: https://www.aoac.org/scientific-solutions/standards-and-official-methods/ (accessed on 23 February 2023).
15. Papotti, G.; Bertelli, D.; Bortolotti, L.; Plessi, M. Chemical and Functional Characterization of Italian Propolis Obtained by Different Harvesting Methods. *J. Agric. Food Chem.* **2012**, *60*, 2852–2862. [CrossRef] [PubMed]

16. Özkök, A.; Keskin, M.; Tanuğur Samancı, A.E.; Yorulmaz Önder, E.; Takma, Ç. Determination of Antioxidant Activity and Phenolic Compounds for Basic Standardization of Turkish Propolis. *Appl. Biol. Chem.* **2021**, *64*, 37. [CrossRef] [PubMed]
17. Escriche, I.; Juan-Borrás, M. Standardizing the Analysis of Phenolic Profile in Propolis. *Food Res. Int.* **2018**, *106*, 834–841. [CrossRef] [PubMed]
18. Chaves, N.; Santiago, A.; Alías, J.C. Quantification of the Antioxidant Activity of Plant Extracts: Analysis of Sensitivity and Hierarchization Based on the Method Used. *Antioxidants* **2020**, *9*, 76. [CrossRef]
19. SAS Institute. *SAS User's Guide: Statistics*, 7th ed.; SAS Institute Inc.: Cary, NC, USA, 1996.
20. El Menyiy, N.; Bakour, M.; El Ghouizi, A.; El Guendouz, S.; Lyoussi, B. Influence of Geographic Origin and Plant Source on Physicochemical Properties, Mineral Content, and Antioxidant and Antibacterial Activities of Moroccan Propolis. *Int. J. Food Sci.* **2021**, *2021*, e5570224. [CrossRef]
21. Park, Y.K.; Alencar, S.M.; Aguiar, C.L. Botanical Origin and Chemical Composition of Brazilian Propolis. *J. Agric. Food Chem.* **2002**, *50*, 2502–2506. [CrossRef]
22. El-Guendouz, S.; Lyoussi, B.; Miguel, M.G.; Figueiredo, A.C. Characterization of Volatiles from Moroccan Propolis Samples. *J. Essent. Oil Res.* **2019**, *31*, 27–33. [CrossRef]
23. Sauri-Duch, E.; Gutiérrez-Canul, C.; Cuevas-Glory, L.F.; Ramón-Canul, L.; Pérez-Pacheco, E.; Moo-Huchin, V.M. Determination of Quality Characteristics, Phenolic Compounds and Antioxidant Activity of Propolis from Southeastern Mexico. *J. Apic. Sci.* **2021**, *65*, 109–122. [CrossRef]
24. Gomes, S.; Dias, L.G.; Moreira, L.L.; Rodrigues, P.; Estevinho, L. Physicochemical, Microbiological and Antimicrobial Properties of Commercial Honeys from Portugal. *Food Chem. Toxicol.* **2010**, *48*, 544–548. [CrossRef]
25. da Silva, J.F.M.; de Souza, M.C.; Matta, S.R.; de Andrade, M.R.; Vidal, F.V.N. Correlation Analysis between Phenolic Levels of Brazilian Propolis Extracts and Their Antimicrobial and Antioxidant Activities. *Food Chem.* **2006**, *99*, 431–435. [CrossRef]
26. Bonvehí, J.S.; Gutiérrez, A.L. The Antimicrobial Effects of Propolis Collected in Different Regions in the Basque Country (Northern Spain). *World J. Microbiol. Biotechnol.* **2012**, *28*, 1351–1358. [CrossRef] [PubMed]
27. Özkırım, A.; Çelemli, Ö.G.; Schiesser, A.; Charistos, L.; Hatjina, F. A Comparison of the Activities of Greek and Turkish Propolis against *Paenibacillus larvae*. *J. Apic. Res.* **2014**, *53*, 528–536. [CrossRef]
28. Kumazawa, S.; Hamasaka, T.; Nakayama, T. Antioxidant Activity of Propolis of Various Geographic Origins. *Food Chem.* **2004**, *84*, 329–339. [CrossRef]
29. Benzie, I.F.; Strain, J.J. The Ferric Reducing Ability of Plasma (FRAP) as a Measure of "Antioxidant Power": The FRAP Assay. *Anal. Biochem.* **1996**, *239*, 70–76. [CrossRef]
30. Martín, I.; Revilla, I.; Vivar-Quintana, A.; Betances Salcedo, E. Pesticide Residues in Propolis from Spain and Chile. An Approach Using near Infrared Spectroscopy. *Talanta* **2017**, *165*, 533–539. [CrossRef]
31. Kasote, D.M.; Pawar, M.V.; Bhatia, R.S.; Nandre, V.S.; Gundu, S.S.; Jagtap, S.D.; Kulkarni, M.V. HPLC, NMR Based Chemical Profiling and Biological Characterisation of Indian Propolis. *Fitoterapia* **2017**, *122*, 52–60. [CrossRef]
32. Sahlan, M.; Rizka Alia Hapsari, N.; Diah Pratami, K.; Cahya Khayrani, A.; Lischer, K.; Alhazmi, A.; Mohammedsaleh, Z.M.; Shater, A.F.; Saleh, F.M.; Alsanie, W.F.; et al. Potential Hepatoprotective Effects of Flavonoids Contained in Propolis from South Sulawesi against Chemotherapy Agents. *Saudi J. Biol. Sci.* **2021**, *28*, 5461–5468. [CrossRef]
33. Socha, R.; Gałkowska, D.; Bugaj, M.; Juszczak, L. Phenolic Composition and Antioxidant Activity of Propolis from Various Regions of Poland. *Nat. Prod. Res.* **2015**, *29*, 416–422. [CrossRef]

MDPI AG
Grosspeteranlage 5
4052 Basel
Switzerland
Tel.: +41 61 683 77 34

Foods Editorial Office
E-mail: foods@mdpi.com
www.mdpi.com/journal/foods